Y0-BTC-268

BORZOI
BOOKS
FIFTY
YEARS

A WINE-GROWER'S GUIDE

A Wine-Grower's Guide

REVISED EDITION (Second)

By Philip M. Wagner

An interesting and informative book for the amateur viticulturist on the cultivation and use of wine grapes

NEW YORK : ALFRED A. KNOPF
1965

L. C. catalog card number: 65-18752

THIS IS A BORZOI BOOK,
PUBLISHED BY ALFRED A. KNOPF, INC.

Manufactured in the United States of America, and distributed by Random House, Inc. Published simultaneously in Toronto, Canada, by Random House of Canada, Limited.

Published September 20, 1945

Second printing, January 1955

Second edition, revised, October 1965

for Jocelyn

PREFACE

THIS BOOK is intended for those who want to grow grapes for wine, and also for those who aren't sure they would care to go to that much trouble but use table wines and would like to know a little bit more about them.

It is a companion to my *American Wines and Wine-Making*. That book has to do with the conversion of good grapes into good wine; this one deals with what comes before that—namely, the production of the good grapes. It is so easily forgotten in this age of man-made wonders that wine is a natural wonder. We can guide the evolution of the grape's juice into wine, but only up to a point. Whether it comes out a fine wine, a mediocre wine, or a barely drinkable wine is beyond our control. That depends on the choice of the right vines in the first place, on the care given them throughout the year but especially during the growing season, and on favors that the climate bestows or withholds. Every experienced wine-maker knows that the only way to be sure of a vintage of the right grapes is to grow them himself. If the novice doesn't start out with that understanding, he soon gets it; and when he does, this book, I hope, will serve as a guide.

Since this is a practical handbook of wine-growing and a layman's survey of the subject, the botanists and ampelographers will find it wanting in many respects. My feeling has been that readers who are approaching the subject for the first time—and those are the readers I have had in mind while writing—will not relish a mass of technical detail. They can always look into such matters later if they want to. But first they want to know what grapes they can grow with hope of success, what kinds of wine these grapes will provide, and the rudiments of the art of growing them. These are the questions I try to answer.

Though the questions remain the same, the answers have changed a good deal since the first edition of this book was published twenty years ago, and a rather thorough overhauling has been necessary. We now know a lot more about grapevines. This is especially true in the East, where some of the new French hybrids are now well established and steadily supplanting some of the old native grapes and where exciting work is being done with even better varieties. The new fungicides and insecticides have enabled us to give better protection to our vineyards. We have a better understanding of plant nutrition and soil management. New tillage implements have displaced the man with the hoe. Though Nature still has the last word, we are not quite so much at her mercy.

It goes without saying that most of this book is unoriginal. It has its basis in our own work with our own ever-changing collection of vines in Boordy Vineyard. But wine-growing is an ancient art and has been developed by the work and study of many men over many years. In some of its essentials the instruction offered by Cato in the oldest Latin text extant does not differ greatly from that given in some of these chapters. No one can write about wine-growing without saying thank you to Marcus Porcius Cato.

From Cato it is a long jump to contemporary wine-growers—in France and Germany and Italy and Portugal, but especially France, whose beautiful vineyards we have explored; and in California, Ohio, Pennsylvania, New York. Wine-growers the world over are members of a club without a name. They have a language of their own that leaps over ordinary linguistic barriers, and what one member knows another is welcome to. I have found this to be true even in Soviet Russia.

I should like to thank especially Professor Maynard A. Amerine and Professor Emeritus A. J. Winkler, of the Department of Viticulture and Enology at the University of California in Davis, whose knowledge is reflected in many parts of this book, and Professor H. P. Olmo, of the same institution; in the East, Mr. Adhemar de Chaunac, of Niagara Falls, Canada, originally from the Cro-Magnon country; Mr. Charles Fournier, Dr. Konstantin Frank, Mr. Seaton C. Mendall, and Mr. Greyton Taylor, all of Hammondsport in the New York Finger Lakes region and all dedicated, though in such different ways, to the same end; Mr. Douglas P. Moorhead, of Moorheadville, North East, Pennsylvania, representing the next generation of American wine-growers; in the South, Mr. R. T. Dunstan, a distinguished grape-breeder; Mr. Frank Schoonmaker, an old friend whose influence on American wine-growing has been pervasive and all to the good; and ever so many more whom I cannot name.

Finally, as the buds swell and another growing season begins, I should like to say that this book is really a testament to the satisfactions that my wife, Jocelyn, *la vigneronne*, and I have found in this vocation, as old as civilization but always new. We like to think that it will show the way to others.

PHILIP M. WAGNER

Boordy Vineyard
Riderwood, Maryland
March 26, 1965

CONTENTS

ILLUSTRATIONS

A WINE-GROWER'S GUIDE

Chapter I

WINE FROM THESE GRAPES

NOT ALL grapes are for the wine vat. Some kinds are made into dried currants and raisins.[1] Others find their way into jars as jelly. Others are only for eating fresh. And some yield up their juices to make an inoffensive beverage called grape juice.

But this book is about wine grapes. So it seems best to begin with a few observations on wine itself, its properties and how it is made, together with some comments on the special characteristics that distinguish wine grapes from those unsuitable for wine-making. This will refresh the mind of the experienced wine-maker and provide bearings for the novice. Then we can move on to such matters as the choice of varieties, pruning, training, methods of culture, and so forth.

Any grapes, if they be sound and ripe, can be made to yield wine of a sort. For the yeasts are not choosy; and fermentation proceeds in the same gaseous frenzy and with perfect impartiality on the wrong grapes as well as the right ones. In either case the resulting fermented liquid satisfies the legal definition of "natural wine." [2] But, alas! not even the majesty of the law can convert a bad wine into a good wine. So it turns out that unless the right grapes were used in the first place, the resulting wine is always disappointing.

[1] Dried currants are not currants at all, but the dried fruit of the grape variety known as Panariti.

[2] "Natural wine within the meaning of this Act shall be deemed to be the product made from the normal alcoholic fermentation of the juice of sound, ripe grapes, without addition or abstraction, except such as may occur in the usual cellar treatment of clarifying and aging. . . ." (Section 610, Act of February 24, 1919, as amended by Section 11, Act of August 29, 1935, and Section 330, Act of June 26, 1936.)

In short, the key to quality lies in the character of the grapes from which wine is made. The number of grape varieties, counting only *wine* varieties, runs into thousands. Hence the endless and fascinating variations, some subtle and some gross, that distinguish wines from one another.

To be sure, other factors influence the character of wine. One of these, and a very important one, is the climate in which a given grape variety is grown. Another, considerably less important than climate, is the soil. Still a third factor is provided by the wine-maker, who, despite the fundamental simplicity of the wine-making process, is capable of affecting the quality of the wine by variations in his technique. Nevertheless – and this is a point that cannot be overemphasized – the variety of grape from which the wine is made is by all odds the most important factor.

Certain grape varieties, no matter how well grown, no matter how attractive they look or how suitable they may be for other purposes, and no matter how ideally adapted to the conditions of a given locality, can never make really good wine. The familiar Concord grape is an example. The Concord grows to perfection in many parts of the eastern United States. But its natural wine, even when made from the peerless Concords of the Chautauqua district of New York State, is a bitter and unpleasant dose, with a rank aroma and a coarse, harsh flavor. By means of elaborate sophistications, and by blending it with wines of other varieties, it can be made to sneak by the noses of the unwary. The fact remains that no one, given a choice, will prefer Concord wine to some other. The conclusive proof is the trivial market that Concord wines, by comparison with the red wines of better varieties, have made for themselves. The Concord grape, handsome, vigorous, and productive, offers a persistent illusion to wine-growers.

Certain varieties, on the other hand, if grown in a region that suits them, may always be counted on to produce good sound agreeable ordinary wines. The grape varieties that dominate the vineyards of California – Alicante Bouschet, Carignane, Mataro, Zinfandel – are representative of this category. In the very best locations they may even produce wine better than ordinary. But no matter how well grown, their wine is never of genuinely superior quality, never "fine wine" in the sense in which wine experts use that term. Experienced wine-growers know this. Such grapes are the bulk-wine varieties, the source of most of the world's humble, agreeable ordinary wine.

Finally, there exist a limited number of grape varieties, most of them rather shy bearers, all of them choosy as to climate and inclined to be particular as to soil, which are the source of the truly fine wines. In these, and in these only, does one find that perfect balance of qualities which, when translated into wine, rewards the wine-maker with something approaching perfection. The names of these varieties are honored among wine-growers, the most famous of all being those called Pinot Noir, Cabernet Sauvignon, Chardonnay, Riesling, and Sémillon. In Europe they are the sources, respectively, of the greatest of the red Burgundies, the Clarets, the white Burgundies, the wines commonly called Rhine and Moselle, and Sauternes. With the exception of Riesling, all of these make fine wines when grown in the proper parts of California. But, like all things of perfection, fine wines even from the finest varieties are rare. No man of modesty and intelligence has a right to expect that, in setting out a vineyard wherever it may suit him, he can make wine to rival the finest. The selection of these varieties and the discovery of those limited areas of the earth's surface where they are happiest were the work of centuries. The finding of new localities perfectly congenial to them must likewise be a tedious process filled with disappointment.

The modest man, however, gains some comfort from the fact that most people have never even tasted one of the "great" wines. Millions look upon wine as an indispensable element in their diet. But the wine they know so well is made from grapes that fall into the middle category — solid, sound wines with a range of aromas and flavors that in itself offers a satisfying lifetime of varied experience. And though

I have ventured to divide wine-grape varieties into three main categories, the fact is that the dividing lines between these categories are very flexible. Actually, though "great" wines are few, the gradations between "great" and "ordinary" are infinite.

At this point it may be useful to offer a digression on "taste" and the critical faculty as applied to wine. People say: "But, after all, wine is a matter of personal taste. How can you say without qualification that such and such a wine is a 'great' wine? Personally, I'd much rather have a glass of sweet Scuppernong than the best bottle of Burgundy you can offer." To such people the analogy of taste in literature may be illuminating. Comic books, for some, are the peak of literary satisfaction. Others find their pleasure in the lachrymose serials of the ladies' magazines. Others are content to curl up with a detective story and consign the rest of literature to oblivion. Above these rudimentary pleasures one enters into the foothills of real literature. Here, reading matter becomes something more than a soporific or a casual distraction. The critical faculties are aroused. The critics have their say; and by a process of interaction between the self-conscious critic and the intelligent reader a gradual winnowing takes place. Finally, in consequence of this complicated process, out of the millions of words that are written every day some few are added to the permanent body of great literature. And when we reach these summits there is no longer room for doubt. The small boy will yawn over Shakespeare, but that is his fault, not Shakespeare's. Similarly, a man may prefer sweet Scuppernong to Richebourg 1959; but that is his fault, not the fault of the Pinot grape, or of the Richebourg vineyard, or of the flawless ripening weather of the Côte d'Or during the autumn of 1959, or of the skillful men who converted those grapes into wine.

Let us cheerfully concede that the best of wine is to be placed among the minor works of art, and that an analogy between "great" wine and great literature is a rather forced one. Let us concede also that most self-styled connoisseurs of wine are not only fakes but dreadful bores. The fact remains that, granting a familiarity with and an interest in wine, persons of average intelligence and normal sensory equipment *do* distinguish degrees of quality in wine – and do so with remarkable agreement. They recognize a good ordinary wine for what it is. They recognize and appreciate wines of better than ordinary quality. And when confronted with a wine that deserves to be called "fine," they need neither label nor sales talk to recognize it.

2

So much for the digression on what may be called the esthetics of wine. It may be taken for granted that the distinctions between wines are real and are recognized and agreed to without difficulty by persons of average intelligence and normal sensory equipment. It follows, too, that practice in wine-tasting heightens the awareness of these distinctions. And we have seen that, fundamentally, and after making due allowance for the influence of climate and soil and the art of the wine-maker, it is the grapes themselves that are responsible for the differences. Now let us see what the grape really is from the wine-maker's point of view. For, though the fruit of no two varieties is identical, the fruit of all has certain characteristics in common.

The Cluster. Grapevines bear their fruit in clusters. According to variety, these clusters may consist of a few berries or many hundreds of berries, and may range in weight from an ounce or two to several pounds. In shape the clusters may be short and broad, or long and narrow. They may be pyramidal or cylindrical. They may be simple – that is, consisting of a single central stalk to which the berries are attached – or they may be compound. In the case of some compound clusters, the stalk branches out to give the cluster a wing, or a shoulder, as in the case of the familiar Delaware grape or the red-wine grape known as Petite Sirah. Or the cluster may be characteristically double shouldered, as in the case of the French hybrid called Seibel 6339. Or it may be many-branched, as in the case of such common California eating grapes as Thompson seedless and Flame

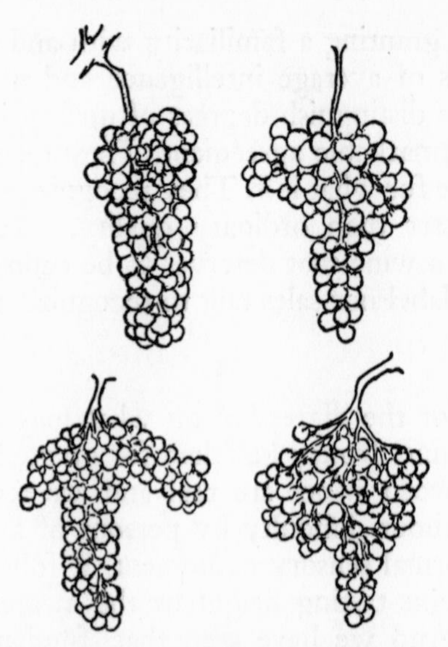

Fig. 1. *Typical cluster shapes: cylindrical, conical, shouldered, compound.*

Tokay. The cluster may be very loose, may be moderately compact, or may consist of berries so tightly packed together that at their bases they assume a hexagonal shape, like the cells of a honeycomb. In some varieties, the stalk becomes lignified as the grapes ripen, so that it must be cut from the vine with a knife; in others, it remains green and succulent to full ripeness, so that the bunches may be snapped off easily. The clusters of some varieties will hang indefinitely on the vine; others will fall when the fruit is overripe. In the case of all varieties, the stalks contain a substantial amount of tannin, a fact which is of considerable importance to the wine-maker. Big clusters mean ease of picking and, usually, big crops; small clusters mean much slower picking and usually smaller crops per vine.

Berries. The berries are attached to the stem of their cluster by means of a pedicel, the organ through which the developing berry gets its nourishment from the vine. The berries of some varieties adhere tightly to the cluster; those of other varieties are subject to "shelling" as they approach ripeness, a defect that requires very prompt picking. The berries range in size from that of blueberries or currants (such varieties as Panariti, Baco No. 1, Léon Millot) to that of plums (many of the showy but relatively tasteless eating grapes); and they have characteristic shapes, of which some are shown in Fig. 2.

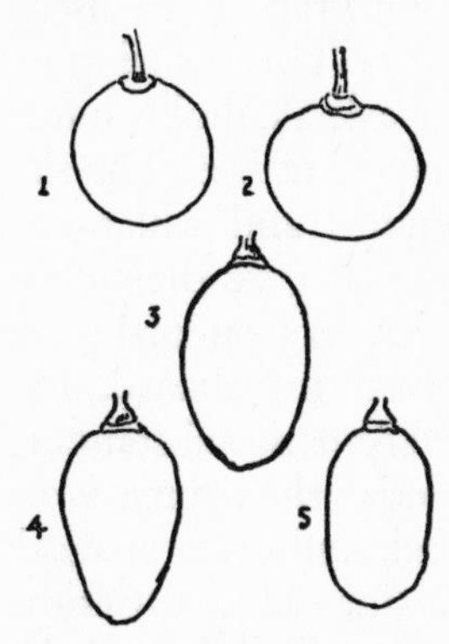

FIG. 2. *Typical berry shapes:* 1, *globular;* 2, *oblate;* 3, *ellipsoidal;* 4, *ovoid;* 5, *elongated.*

The berries consist of *skin*, *seeds*, and *pulp*.

The skin has great importance to the wine-maker. That of most varieties is covered with a waxy bloom, in some cases so abundant that it seems almost like powder. This characteristic is expressed in the names of some varieties—for example, that of the minor French Burgundy variety, Meunier.[1] It is this outer bloom that imparts a kind of pearly iridescence to certain white varieties, and gives a handsome brightness to the blue of many others. This bloom has certain practical functions: it sheds water easily; apparently it protects the grapes from "sunburn"; and it has the function of catching and holding the spores of yeast, so that yeast is always present on the grapes in abundance when the time for fermentation arrives.

The skin also contains, in its inner layers, the coloring matter of the berry. In some varieties this coloring matter

[1] Meaning "miller."

is intense and abundant; in others, it is present in small quantities only, so that the grapes appear pink or rose colored rather than blue or black. In still other varieties — the so-called white grapes — it is almost but not entirely lacking. This coloring matter is practically insoluble in water at ordinary temperatures, but not at high temperatures — a characteristic that the grape-juice users take advantage of in extracting the color.[1] This coloring matter is highly soluble, however, in a weak alcoholic solution — a characteristic that the wine-makers utilize. The skin is also the seat of the numerous, complex, and still only partly understood aromatic substances that give their characteristic aroma to certain grape varieties. In Concord grapes, in the Scuppernongs of the southeastern seaboard, and in varieties of the muscat family these aromatic substances are extremely powerful. In other varieties the aroma is much more delicate, but no less individual, and gives attractive character to the wine. Most of the fine wine varieties — Pinot, Cabernet, Riesling, and so on — are mildly and distinctively aromatic grapes. Or the aroma may be so slight as to be practically unnoticeable. This is characteristic of the greater number of wine-grape varieties.

It may be worth while at this point to make the distinction between *aroma* and *bouquet.* Aroma is carried over directly into the wine from the fresh grapes. Bouquet is a quality which young wine always lacks, and is derived from a slow and complicated chemical reaction that takes place in the wine during the period of aging. Aroma tends to disappear as the wine ages, whereas bouquet tends to increase in intensity.

As the grapes ripen, their aroma is progressively more noticeable, being imparted at full ripeness into the pulp and through the skin into air. An experienced wine-maker is very attentive to this point. In the case of aromas that improve the quality of the wine, he waits for their full development. On the other hand, in the case of grapes that possess

[1] They extract the color by crushing the grapes and heating them before pressing out the juice.

an aroma unattractive in wine, he can reduce it somewhat by picking just prior to full ripeness.

The skin also contains tannin in considerable quantities, a substance that promotes a clean and healthy fermentation, helps to stabilize the color of a red wine, and gives an agreeable astringency to certain wines, such as the Italian Chianti and some of the Clarets. Finally, the skin contains numerous other substances in very slight amounts; and some of these, despite the very small amounts present, play an important role in determining the character of the wine.

The seeds are present in the berry in numbers ranging from one to six, according to variety; and even the seedless varieties have seeds, though they are small, soft, defective and nonviable. The seeds have an interest only for the grape-breeder intent on producing new varieties, and for the ampelographer, who finds in their shape, size, and color some clues as to the parentage of the varieties he is trying to identify. So far as the wine-maker is concerned, he is chiefly interested in seeing that the seeds are not actually broken when the grapes are crushed, since the bitter protein matter inside the tough casing of the seed can impart bitterness to the wine. The seeds also contain much tannin, but from the wine-maker's point of view the tannin in grape skins and stems is much more accessible.

The pulp is the principal part of all cultivated varieties of grape, though an unimportant part in most wild grapes.[1] In the cultivated varieties, the pulp constitutes approximately 95% of the berry; and the pulp in turn consists of water (75–80%) and sugar (15–25%). But in addition to these it contains appreciable quantities of acid salts and free acids, of aromatic compounds, and of other ingredients such as pectin — all of which are immensely important in determining the character of the finished wine. As we shall see, the factors of water, sugar, and "total acidity" are especially

[1] From Nature's point of view, the main thing about a grape berry is its quota of seeds, since these insure the continuation of the species. The skin is merely a protection to the seeds; and as for the pulp, it is without importance, save perhaps in inducing some birds to eat the berries and thus leading to the wider dissemination of the species.

important if the wine is to be well balanced and stable. The pulp of some varieties (the California eating varieties, for example) is adherent to the tender skin, so that the two are eaten together. The pulp of many of our eastern varieties, such as the Concord, slips easily from the skin, giving rise to the term "slip-skin grapes." The consistency of pulp is tough in some varieties, which causes trouble for the wine-maker; in other cases it may be crisp but tender, or very liquid. The pulp of most grape varieties is colorless, since (as we have seen) the color of blue and red grapes is contained in the skin. However, there exist a limited number of varieties that have coloring matter in the pulp as well as in the skin. These varieties – such as the Alicante Bouschet, extensively grown in California – are known by the term *teinturier*, since their wines are more deeply colored than most and are used for blending with wines that lack sufficient color.

3

To summarize: the raw material from which wine is made consists of the skins, pulp, seeds, and stems of grapes, which vary greatly in their proportions and in their composition according to variety, degree of ripeness, and the influences of soil and climate. Now let us see what the wine-maker does with these raw materials. A general view of the art of the wine-maker,[1] bent on extracting the utmost in quality from his materials, ought to be taken if we are later to judge the relative virtues of the many varieties of grapes.

Red Wine. To make red wine, grapes containing ample pigment (that is, blue or black grapes) have to be used. The fruit is crushed into an open vat, barrel, or large crock. It may be crushed by foot power or by being run through a crusher, which roughly resembles the wringer of a washing machine and is operated by hand or by power. As soon as the grapes have been crushed, the wine-maker dips out a

[1] What follows is only a synopsis. For a full discussion of wine-making see Philip M. Wagner: *American Wines and Wine-Making* (New York: Alfred A. Knopf; 1963).

sample of the juice, which is colorless or at most slightly pink.

He tests this juice for its sugar content with a simple little floating device called a saccharometer. Ideally, the sugar content is between 20 and 24% by volume, which may be expected to yield a wine containing 10–12% of alcohol by volume.[1] If the sugar content turns out to be less than 20%, as it frequently is in certain varieties grown in many parts of the United States, he adds enough ordinary granulated cane sugar to bring the sugar content of the juice up to 20–24%. This is entirely legal and is practiced in every wine-growing district of the world, the reason being that juice of (say) 16% sugar yields a wine containing only 8% of alcohol; and that is insufficient for stability and keeping quality in the wine. If on the other hand the wine-maker is working with the wine grapes of California, he may find that the sugar content is above 24%. In that case, if he is wise, he will blend the juice with juice of other grapes having a lower sugar content, so that the average falls within the proper limits. Otherwise he is likely to have trouble in the course of fermentation, since fermentation proceeds with difficulty in too high a concentration of sugar.

Having tested the sugar content, the wine-maker then measures the "total acidity." This is likewise a very simple and rapid test, though not quite so simple and rapid as the test for sugar. Ideally, the acidity falls between .6 parts per 100 and 1. per 100. If the acidity falls below .6, he runs a serious risk of trouble during fermentation, for the agents of disease thrive in the absence of acidity. And, even though the fermentation proceeds without event, the lack of acidity will cause the wine to taste flat, heavy, and lifeless. If on the other hand the acidity is very much above 1., then the wine will be too tart, too "green." If the wine-maker is working with the wine grapes of California, he will find that the acidity is frequently below the ideal, since low acid is usually combined with high sugar. If he is working with

[1] Two per cent of sugar yields approximately 1° of alcohol in the finished wine.

the wine grapes grown in the East, he will frequently find that the acidity is too high, since high acidity is usually found with low sugar. In either case he makes adjustments. For low acidity, he adds an appropriate quantity of tartaric or citric acid. Excessive acidity he reduces by the addition of a small quantity of water – though this is a procedure which he resorts to only under dire necessity. Other ways of reducing the acidity are preferable, the chief of these being, of course, the blending of high-acid wine with low-acid wine.

Assume that the juice, or "must," has been properly balanced as to sugar content and total acidity. Now it is ready for the fermentation. The traditional method is to do nothing about this, merely counting on the yeast that was present on the skins of the grapes. This method is still largely used in regions where the grapes have ample acidity, there being little risk that the "bad" ferments always present along with the "good" ferments will get the upper hand. However, wine-makers in low-acid regions such as California play safe by adding a "starter" of violently fermenting yeast to the newly crushed grapes, thus assuring the predominance of the right yeast.

Once the fermentation is under way, there is no room for doubt of it. Carbon dioxide gas, with a characteristic pungent and delicious odor, is thrown off in large quantities; the mass of pulp and skins begins to disintegrate, the skins rising to the top of the mass with the free juice beneath. Twice a day or more the wine-maker punches this "cap" of skins and stems down into the juice, to insure full extraction of the coloring matter and a mild aeration of the fermenting mass.

The violent fermentation for red wine lasts from three to eight or ten days, depending on the temperature at which it proceeds. In the course of it, the color is extracted from the skins, and appreciable quantities of tannin, aromatics, and other substances are extracted from skins and stems. Frequently the wine-maker may judge that the extraction of these elements has gone far enough before the fermentation is finished. In that case he draws off the partly fermented juice into a proper container, presses the soggy mass of skins and stems for whatever juice remains in them, and lets the fermentation continue to conclusion in the absence of skin and stems. The character of the wine may be considerably altered by the length of fermentation "on the skins." In making what is called *rosé* (pink) wine, the wine-maker may allow the fermentation to proceed on the skins for as short a time as 18 hours, in order to obtain only a partial extraction of color, tannin, and so on.

In any case, the young wine must be drained off, and the "marc" – or pomace, as it is usually called in American wineries – is pressed as soon as the violent fermentation has subsided, the free-run wine and the press wine usually being put together. This young wine is muddy stuff, and unpleasantly yeasty and harsh to the taste. The wine-maker fills his containers *full*. Not only that: he carefully replaces any loss by evaporation or by contraction caused by cold weather, to avoid leaving any air space in the container. In the following months the sediment falls, and the wine gradually grows bright and clear. In the course of the first year it is "racked" from the sediment, or lees; and after a length of time ranging from six months to two years (depending on the character of the wine and hence chiefly on the character of the grapes) it is ready to bottle and drink. When it is young, the wine is frequently fresh and "fruity"; as it grows older the fruitiness is displaced by bouquet.

White Wine. White wine may be made from white grapes, pink grapes, or even red-wine grapes. The fundamental distinction between red wine and white wine is that the latter is made from juice which is fermented by itself and not in mixture with the skins and stems. Since the pigment is contained in the skins, the juice may therefore be pale in color even though the pigment in the skins is abundant.

To make white wine, the wine-maker therefore crushes the grapes and then presses them immediately. The ease of pressing depends to a considerable extent on the character of the grape, those with tough pulp, such as the Catawba, being hard to press, and those with a liquid pulp such as Sémillon or Seibel 5279 being relatively easy to press. If red-wine grapes are used, the pressing must be not only very rapid but gentle as well; otherwise a certain quantity of the pigment will be present in the juice and the resulting wine will be what the French call *taché* – that is, neither white nor pink but an unattractive shade of pinkish brown.

Once the pressing is done, the wine-maker tests his juice for sugar and acidity (as in the case of the must for red wine) and makes any necessary corrections in its balance. The juice, instead of fermenting in vats as for red wine, goes through this operation in casks that are filled not more than two-thirds full. The fermentation starts more slowly than that for red wine and never attains the same violence. To ferment out fully, a longer time is also required – not infrequently as long as two months. When the fermentation has finally ceased, the wine is siphoned from its gross sediment and put into clean containers, which are kept full, and is treated henceforth very much as new red wine is treated. However, white wine, owing to its different composition, is usually lighter in body, fresher tasting, more delicate, and ready to drink sooner, than red wine. And, more often than red wine, it depends for its attractiveness on the aroma of the grape rather than on the subsequent development of bouquet.

Special Wines. The foregoing synopsis of wine-making has dealt with the making of light dry table wines only. The making of sparkling wines, fortified wines, natural sweet wines, and other specialties is a much more complicated business, and is not within the scope of this book.

From this synopsis, it must be evident that the wine-maker is chiefly concerned with the balance of the principal ingredients: sugar, acidity, and water – in short, with those characteristics which insure a "sound" wine, a wine that will be stable and healthy – and with insuring a normal fermentation. As for the overtones – the influence of the minor yet very important ingredients, the characteristics that make the difference between a wine that is sound and nothing more and a wine of superior quality — there is very little that he can do. His function in the natural process may be likened to the role of the baby doctor. The doctor can deal with disease; he can sometimes correct abnormalities; he can often *prevent* disease. In a word, he can insure physical "soundness." But when it comes to those subtle and little understood qualities of the human organism which produce sometimes a genius and sometimes a dullard, there is very little he can do. Similarly, when confronted with the "personality" of his grapes, the wine-maker is largely helpless. At most, he can confine his attentions to those grapes which, experience tells him, have promising wine personalities, and leave the others to the fruit stalls and the jelly manufacturers.

Chapter II

GRAPEVINES, IMMIGRANT AND NATIVE

ONE MIGHT perhaps assume, from what was said in the previous chapter, that the intending wine-grower has only to decide what kind of wine he wants to make, then plant and tend vines of the appropriate variety, and from that point on enjoy a plenitude of his favorite wine. The function of this chapter is to dissipate – or at least to qualify – that agreeable assumption.

For the fact is that grape varieties differ greatly in their cultural requirements. Some like it cool, and some like it hot. Some like it dry, and some like it humid. Some like sand, and some like clay. Some can tolerate lime and others perish if the lime content of the soil is above a certain point. As has already been stated, there are hundreds, even thousands, of grape varieties. What has not been said is that there are many species as well as varieties. The genus *Vitis* of the Family *Ampelidaceae* has representatives throughout the temperate and subtropical world. The Chinese have their cultivated grapes, and there are species that grow wild on the river banks of northeastern Siberia. Europe has its grapes both wild and cultivated. Central America and the Caribbean have their species, as yet little studied. North America is particularly rich in species, their number running into the dozens.[1] These numerous species have evolved under highly

[1] The exact number depends on the botanical classification that one chooses to follow. Classifications into species are after all quite arbitrary, and subjective elements are not easily ruled out, since the ease with which grapes interbreed gives rise to innumerable borderline subjects not readily classified.

varied conditions, and they are not always adaptable to new environments.

Of these, one species in particular dominates the history of grape culture. This is *Vitis vinifera*, to which all of the classic wine grapes of the Old World are assigned – and all the Old World table grapes, too, for that matter. Its origin seems to have been in the Middle East, in the region of the Caucasus mountains between the Black and the Caspian seas – perhaps, as Genesis seems to suggest, on the very slopes of Mount Ararat. However that may be, *Vitis vinifera* has prospered mightily under the hand of man, who has insured the survival not only of the fittest but of the finest. By a process of selection, partly natural and partly artificial, its seedlings have provided an immense assortment of varieties. These varieties differ not only in the character of their fruit but in character of vine – so much so that an adequate catalogue of all their differences would require many volumes. By a process of diffusion combined with continued selection, local members of the great *vinifera* species are to be found all through the Middle East, the Near East, and that luxuriant seedbed of cultures, human as well as botanical, the Mediterranean world. The species spread eastward, too, through southern Asia. And from the Mediterranean it worked its way gradually northward into Europe to the present limit of vine-growing in France and Germany. From Europe the best of its varieties, whether for wine or for fresh fruit, have spread far. Today they are the basis of wine-growing in the great regions of North Africa, South Africa, Australia, California, Chile, and Argentina.

But, despite their immense diffusion, the vines of the species *Vitis vinifera* have their definite limits, beyond which they do not go. Though their vines are grown in most parts of France, none will be found in Normandy and along the English Channel. Nor are the vines grown across the Channel in England, except in hot-houses; nor in neighboring Belgium and Holland. Likewise, though they flourish in Argentina on the eastern slopes of the Andes, they will not succeed in neighboring Brazil. And though they thrive in California, they have no taste for that greater part of North America which lies east of the Rockies.

2

Would-be wine growers in the United States have learned this the hard and bitter way, in the course of a good three centuries of unsuccessful effort to grow them. And still, despite the accumulated evidence of the three centuries, new generations of would-be wine-growers, if they be of the inquiring or skeptical kind, usually add their own to the accumulation of disappointment and adverse evidence.[1]

There is no need to repeat the detailed history of American misadventures with the *vinifera* grapes; that has been done before and by more competent hands.[2] A few random instances will suffice.

Chief inspiration of the false hopes of the earlier experimenters was the presence, all along the Atlantic seaboard, of wild grapevines in great profusion. Their fruits had unfamiliar flavors, and their vines were rather dismal substitutes for those whose memory the earlier settlers brought with them. Edward Everett Hale recorded that the Massachusetts colonists made wine of the native grapes at the end of their first summer but that (with admirable understatement) "the appetite for such wine does not seem perilous"; and similarly tepid enthusiasm for the wines of the native wild grapes was recorded by men of the other colonies.

These pioneers reasoned, however, that if wild vines grew in such luxuriance, the tame cultivated varieties should do

[1] At Boordy Vineyard we have in the course of the years tried a great many varieties of *vinifera grapes*, both grafted and on their own roots, with a total lack of success in the case of most *vinifera* varieties. A few of the northern European *vinifera* survive on appropriate grafted roots and occasionally, with attentive care, these produce a crop and give us some very superior wine.

[2] L. H. Bailey: *Sketch of the Evolution of Our Native Fruits* (New York, 1912); and U. P. Hedrick: *The Grapes of New York* (Albany, 1908).

as well or even better. They sent home for vines and French *vignerons*. They obtained grants for the specific purpose of setting out wine vineyards. In 1632 Governor Winthrop was granted what is now called Governor's Island in Boston Harbor on condition that he plant it to vines; and the rent was fixed at a hogshead of wine a year. As early as 1621, the London Company was arranging to settle sixty families of "ffrenchmen and Walloones" in Virginia, chiefly for the purpose of wine-growing; and a selection of vines was sent over by the ship *Warwick*, Capt. Tho. Nuce, in that year. They reported, on seeing the land, "that it far excell'd their own Country of *Languedoc*." [1] Even before Governor Winthrop's experiment in Boston Harbor, a vineyard of European vines was planted at the mouth of the Piscataqua River near what is now Portsmouth, New Hampshire. The Huguenots, coming to the Carolinas in the late 17th and 18th centuries, brought vines with them. One of the most ambitious of these early wine-growing schemes was that undertaken at Savannah, Georgia, in the 1730's by Abraham de Lyon, a Portuguese, with the encouragement of the colony. But, despite the scale of the experiment and the enthusiasm of Senhor de Lyon, a chronicler of a few years later reported that the vineyard "which was to supply all the plantations . . . resulted in only a few gallons, and was then abandoned."

Consider the case of Thomas Jefferson. Among the many interests of that extraordinary man was gardening, and especially the enrichment of our agriculture by the importation of new plants. He maintained a vast correspondence with botanists, seedsmen, nurserymen, the directors of public gardens, gentlemen farmers, explorers, and humble growers relating to plant and seed introductions. Not only that: he found time to plant a little of everything at Monticello; and from 1766 to 1826 he kept a "garden book" in which he listed [2] his various plantings and schemes for plant-

[1] Robert Beverley: *History of Virginia* (1722).
[2] With many lacunae.

ing. Through all this material runs his continuing interest in the vine. There were times when he had his doubts of the future of wine-growing in the United States. But on the whole he was sanguine:

> Wine being among the earliest luxuries in which we indulge ourselves, it is desirable it should be made here and we have every soil, aspect & climate of the best wine countries, . . . the Italian Mazzei, who came here to make wine, fixed on these South West mountains, having a S.E. aspect, and abundance of lean & meagre spots of stony & red soil, without sand, resembling extremely the Cote of Burgundy from Chambertin to Monrachet, where the famous wines of Burgundy are made. . . .[1]

Numerous records of vine plantings on his own land are recorded in his garden book. In the spring of 1802 (to choose an entry at random) he was planting "30 plants of vines from Burgundy and Champagne with roots, 30 plants of vines of Bordeaux with roots, 10 plants of vines from Cape of good hope with roots." A few years later, in 1807, he set down a long list of grapes recently planted, including such famous *vinifera* varieties as Black Hamburg, Frontignac, Chasselas, Trebbiano (a white-wine variety whose history has been traced back to the Romans), San Giovetto (chief of the Chianti varieties), and Aleatico (grown in Italy for a sweet red Muscatel wine).

But did these experiments yield him wine? In 1786 he was writing from France to Anthony Giannini, an Italian ex-*vigneron* who worked for him intermittently for many years: ". . . how does my vineyard come on? have there been grapes enough to make a trial of wine? if there should be, I should be glad to receive here a few bottles of the wine." Giannini sent him no wine. Nor, so far as the records show, did Jefferson ever have a vintage at Monticello, though there was always plenty of cider and an abundance of peach mobby. A Swiss visitor to Monticello in 1799 found, of the vines that had been planted up to that time, that they "had been abandoned . . . which proved, evi-

[1] Jefferson to John Dortie, Monticello, Oct. 1, 1811.

dently that it had not been profitable." [1] As much was true of the experimental plantings that continued almost up to his death.

The record, then, from the earliest days, was dishearteningly consistent. The vines of Europe, transplanted to the colonies, might grow and flourish for a season, or two, giving rise to extravagant hopes. But time after time, just as they were about to come into full bearing, they failed. The fruits grew black and then shriveled into hard wrinkled mummies. The foliage sickened, withered, and fell from the plants before the fruit could ripen. Then a cold winter would kill the vines to the ground; and finally, resisting all efforts to nurse them along, they would die. Over and over again it happened. We know now, but they didn't then, that the vines of *Vitis vinifera* can never be wholly at home when they face the conditions of high humidity, the indigenous vine diseases, and the cold winters that prevail throughout so much of North America.

3

It was only with the settlement of California that *Vitis vinifera* finally found a congenial home on this continent. California's thousands of acres of vineyards constitute our direct link with the wines and vineyards of the Old World; and the story of the establishment and growth of California's viticulture is convincing positive evidence, as the record of failures in the East should be convincing negative evidence, that Nature has the last say in such matters.

California's viticulture was established by the mission

[1] A similar report was provided by Isaac Weld, Jr. in his *Travels Through the States of North America, and the provinces of upper and lower Canada, during the years 1795, 1796, and 1797:* "Several attempts have been made in this neighbourhood to bring the manufacture of wine to perfection; none of them however have succeeded to the wish of the parties. A set of gentlemen once went to the expence even of getting six Italians over for the purpose, but the vines which the Italians found growing here, were different, as well as the soil, from what they had been in the habit of cultivating, and they were not much [more] successful in the business than the people of the country."

friars incidental to their pious task of conquest. The Spanish friars were naturally anxious to establish the vine and make wine in the New World not only because wine was one of their customary foods but because of its place in the symbolism of their religion. So the vine followed close on their heels. It was in the last quarter of the 17th century that the Jesuits finally crossed the Gulf of California from the mainland of Mexico and established their first mission on the peninsula of Baja California.[1] The wild grapes that the devout and thirsty friars found there[2] were worthless for wine, and they had to satisfy themselves with such small supplies as could be brought over from the mainland of Mexico. But before very long, in 1687, a strong-minded and resourceful priest arrived in Baja California from the Jesuit College in Mexico City. This Fr. Juan Ugarte brought with him the seeds of the usual vegetables, together with some vine cuttings. The record shows that not many years after his arrival he had a thrifty vineyard from which he made wine first for his own mission and later for all the others of Baja California.

The mission system expanded at a fairly rapid rate; and as each new mission was formed by a division from its predecessor, so it obtained cuttings of vines from the vineyards of its parent and tried to make itself self-sufficient so far as wine was concerned. It was by this gradual process that toward the end of the 18th century the vine finally reached California proper, or Alta California. Legend has it that Fr. Junipero Serra brought the first vines from Baja California and planted them at San Diego in 1769. But there is no definite proof of this, and indeed it was another California mission, San Gabriel, which came to be known as the Viña Madre, or mother-vineyard, of all the California mission vineyards. What we do know, though, is that the mission fathers concentrated all their attention on a single variety of vine, a variety that is still grown in California to some

[1] I am indebted to *The Early History of Wine Production in California*, by Herbert Boynton Leggett (Berkeley: University of California Press; 1940) for some of what follows.

[2] Of the species *Vitis californica*.

extent and is now called by the highly appropriate, if not very original, name of Mission.

Why did the mission fathers cultivate this single variety to the exclusion of all others? The Mission variety makes a wine of very mediocre quality at best. Why did they make no effort to try out the viticultural possibilities of the new land more extensively and systematically? Probably the answer lies in the fact that the Fathers dealt very little in wine as an article of commerce, being content to produce merely enough for their own purposes, for the use of the sick, and for those few missions which were unsuccessful in establishing vineyards of their own.

But there is something even more curious about this devotion to a single variety. It is obviously a variety of *Vitis vinifera*, the species to which all the wine grapes of Europe belong. Yet nowhere else in the world has this particular *vinifera* variety ever been encountered. The presumption is that it came from Spain, where most of their other imported supplies came from. Yet Spain knows no grape conforming to its description. One hypothesis which so far as I know has never been offered before, and which seems to me to explain the mystery as well as any, is that the early fathers imported from Spain not only cuttings for vegetative reproduction but seeds. The voyage was a long and hazardous one. It is conceivable that the cuttings may not have survived and that the Mission variety sprang from a grape seed imported from Spain, and, being adequate to the purpose even though not remarkable, was multiplied. Since the seeds of *vinifera* grapes rarely, if ever, breed true to their parent, the absence of an exact counterpart of the Mission in Europe would thus be explained without prejudice to the purity of its *vinifera* ancestry. According to this hypothesis, the Mission is, paradoxically, a European wine grape that is also a native American.

But the question is after all an academic one. When one reads of the wine-making methods that prevailed in most of the missions, the wonder is not so much that they were satisfied with a single mediocre variety of wine grape as that they got any wine at all. Few missions had proper wine presses, and few had adequate cooperage or other wine-making equipment. Thus at one mission, Santa Gertrudis, the German Jesuit Jorge Retz solved the problem by hollowing out some huge rocks which he used first as fermenting vats, and then, by filling with new wine and sealing with pitch, as "casks." Ordinarily the crushing of the grapes was done by "well-washed Indians," their hair tied up and wearing only a *zapata*, whose hands were covered with cloths to wipe off the perspiration, and who thrashed about in the mass of fresh grapes until these were sufficiently macerated, the fresh must being then caught and, with the help of God, fermented into wine in big leather bags.

With the collapse of the mission system in the 1830's, viticulture west of the Rockies entered a new phase. First in the vicinity of Los Angeles and then farther north, private individuals began to undertake grape-growing and wine-making. And unlike the mission fathers, they were not content with the use of a single mediocre variety. It was in 1831 that a Frenchman bearing the portentous name of Louis Vignes arrived in Los Angeles and embarked upon a wine-making enterprise. Vignes, a true Bordelais, had little respect for the grape that he found dominating the region, and succeeded in bringing in small quantities of cuttings of French wine varieties. They thrived. Soon Vignes had no fewer than 104 acres under vines, and not many years later he was supposed to have the largest vineyard in California — though just how large it was is not clear.

Vignes appears, then, to have been the first to experiment seriously with the introduction of other and better *vinifera* grapes from Europe. Most of those who rushed after him into grape-growing and wine-making had no viticultural or wine-making background whatever and indeed no qualifications except an eagerness to get in on a good thing. But they were sufficient in numbers to make wine a cheap staple of the region directly around Los Angeles; and in the decade of the 1850's grape-growing began to be taken up in the northern counties of California as well. These first northern grape-growers, like most of the southern, were speculators. As Charles Kohler [1] remarked: "Nearly every land owner caught the wine fever, entertaining the idea that the planting of a few thousand vines would make him rich." And a contemporary (who evidently had a bad case of the fever) declared that "capital put into vineyards would bring greater returns than when outlayed in fluming rivers for golden treasures."

As these enthusiasts gained some little experience they began to realize what old Vignes had known before he set out a single vine, namely, that if the most was to be made of California's viticultural possibilities, if California was ever to gain a reputation as a source of good wines, grapes better than the Mission would have to be found. And thanks to the efforts of a few of these men [2] most of the important European varieties found their way to California. But there existed no mechanism by which these numerous varieties could be tested systematically for their adaptability to the various parts of the State. And so, from mid-century well into the 1880's, the selection of the more suitable varieties and the discovery of the most favored grape-growing areas proceeded by hit or miss, and in utter confusion. Eventually, however, a great series of experiments into the suitability of various grape varieties was begun by E. W. Hilgard, first director of the State's agricultural experiment station. Hilgard was a scientist of exceptional gifts. Moreover, he was fond of wine. And this wedding of science and passion produced exceptional results. Taken by and large, and considering the limitations under which he worked, Hilgard's conclusions as to the merits and drawbacks of various grape varieties, and as to the possibilities of California wine-growing, were remarkably accurate.

Hilgard was succeeded in turn around the end of the century, as the "dean" of California viticulture, by Professor F. T. Bioletti, who lived through Prohibition. Others, whose

[1] One of the northern grape-growing and wine-making pioneers.

[2] Of whom a remarkable Hungarian, Col. Agaston Haraszthy, was the most important.

names will be encountered in later chapters, are carrying on in the fine tradition established by those two men. We shall see how their labor and knowledge at last have brought order out of confusion.

Throughout its turbulent history, California's viticulture has had a succession of ups and downs. It has been a source of great wealth. But it has also been victimized by economic forces beyond its control, by overconfidence, by fraudulence and irresponsibility, by ignorance, and by the fantastic interlude of national prohibition.

What really matters, though, at least so far as the thesis of this chapter is concerned, is that the troubles of California's wine-growers have been of a wholly different order from those encountered in the East. For the climate of California proved congenial to the species *Vitis vinifera* from that day in the late 17th century when the first cutting was planted by the learned Fr. Juan Ugarte.

Chapter III

VITICULTURAL MELTING POT

1. THE AMERICAN HYBRIDS

THUS WE SEE that the *vinifera* grape has been most reluctant to grow in most parts of the United States, but that from the very day of its introduction into California it has flourished prodigiously. Clearly the viticulture of California will always be based on the *vinifera* grapes.

But what of the East? Though the *vinifera* grapes remain exotics we do know that grapes are grown, in some degree, in every State of the Union, and that in parts of the country other than California drinkable wines have long been made (though in relatively small quantities). What grapes are these that grow where the *vinifera* grapes offer such difficulties?

To understand this, recall the remark made at the beginning of Chapter II that *Vitis vinifera* is only one of many species of the genus *Vitis*. North America is particularly rich in its own species. In the woods and thickets of New England – and spreading, indeed, over a vast territory extending from Canada throughout the entire Appalachian region and the valleys of the western Appalachian slope – wild grapes of the species *Vitis labrusca* flourish exceedingly. These are the so-called "fox grapes." The grapes of another wild species, *Vitis aestivalis*, inhabit a region fully as large, consisting roughly of the central and southern Atlantic coast and the Mississippi Valley. Another species, *Vitis Lincecumii*, which resembles *aestivalis* in some respects, occupies the territory west of the Mississippi from central Missouri south through Oklahoma and Arkansas, and into northwest Louisiana and northeast Texas. There is *Vitis candicans,* native of the Southwest, which has the characteristic, missing in most American grape species, of thriving on land that is rich in lime. Occupying much of the territory congenial to *Vitis labrusca* are vines of another species, *Vitis riparia,* with fruit of an entirely different character and a liking for river banks and other well-watered places. There is, too, the species *rupestris,* to be found in many parts of the Appalachian system and in the hilly lands west of the Mississippi – a species that, in contrast to *Vitis riparia,* does not care at all for moist locations but seems happiest when growing (in a bushy, shrubby way) on meager, rocky, hilly, or mountainous land.[1]

When the first settlers came, they could hardly miss being impressed, as I have suggested already, by these wild grapes. And though most of them reasoned, wrongly, that the cultivated grapes of Europe would surely succeed where wild vines did so well, others concentrated on the wild vines. Though the fruit of these left much to be desired, either for eating or for wine-making, at least the vines could be counted on to produce crops. From the very beginning there was a small minority among the settlers who held that the wise course must be to seek out the very best of the wild vines, bring them into cultivation, and on the basis of them to create a new viticulture.

On page 22 there is mentioned a list of the grapes that Thomas Jefferson planted in his vineyard at Monticello. Among these were "10 vines of the Cape of good hope grape"; and the history of this particular grape illustrates very well the accidental beginnings of our distinctively American viticulture. Jefferson obtained these vines from Peter Legaux of Philadelphia. Jefferson was under the impression that it was in fact a variety of *Vitis vinifera* from the Cape of Good Hope; and so, as far as we know, was Peter Legaux, the man who sold them to him. Actually the Cape grape (it came to be known also as the Schuylkill, Clif-

[1] See table on pp. 200–1 for notes on the characteristics of these and other species that have proved valuable in breeding new varieties.

ton's Constantia, and the Alexander) was not a *vinifera* grape at all, but an offshoot of the wild species *Vitis labrusca*, an accidental seedling, possibly a cross between some wild vine and a vine from Europe that someone was attempting to cultivate near by. At any rate, Mr. Legaux somehow obtained cuttings of this variety and propagated it, retailing it as a European grape. Since Jefferson did not pursue his experiment with vines consistently, we have no record of the behavior of the Cape grape with him. However, the Cape fell into other hands and especially (for the purposes of this brief account) into the hands of one John James Dufour, a Swiss who was developing an ambitious project for wine-growing in Kentucky. With him it became the basis, first in Kentucky and then across the Ohio, for a substantial wine-growing enterprise. As a red wine, it must have been heavily foxy and of inferior quality, but at least it was drinkable; and as a white wine it appears to have been rather better. On this we have the word of Nicholas Longworth (1782–1863), who did not disdain to grow it.

Thus by the beginning of the 19th century this grape was providing evidence of the possibilities of our native species. However, its faults were many; and a rival, the Catawba, was not long in appearing. Credit for the introduction of this grape, which is still more widely grown for white wine than any other grape of American parentage, belongs to a remarkable man. Major John Adlum, an enthusiastic fruit-grower whose home, "The Vineyard," occupied what is now a part of Rock Creek Park in Washington, D. C., found this vine on the premises of a Mrs. Scholl, who kept an inn at Clarksburg in Montgomery County, Maryland; and Mrs. Scholl in turn seems to have acquired it indirectly from North Carolina, where it had been found growing wild in the woods of Buncombe County not far from the present Asheville. But it was Nicholas Longworth, of Cincinnati, who made it famous. Longworth's sparkling Catawba was for many years the standard of quality in American wine; and thanks to his efforts the Catawba grape was spread far and wide.

The success of the Catawba inspired a search for other varieties containing native blood, one of these being the Isabella, a blue grape, which was introduced by the Long Island nurseryman, William Robert Prince. This was also an extremely foxy *labrusca* seedling; but like the Catawba it yielded a satisfactory white wine when pressed and fermented off the skin. The Isabella has since spread far and wide, being grown in places as remote as the more inclement sections of the Balkans, Brazil, and South Africa (where it is used as a rootstock).[1]

Thanks then to the Cape, the Catawba, and the Isabella, forward-looking horticulturists had become convinced by the middle of the 19th century that the native grapes had real importance. It was in the 1840's that the wine-growing region of the Finger Lakes in central New York State got its start, based on the Catawba and the Isabella. Grapes began to expand in the Hudson River Valley; and, too, in Missouri along the banks of the Missouri River. Simultaneously the search for other, and better, seedlings continued. In Missouri, for example, the Catawba did not do well; but a blue grape called Norton's Virginia (the name is now generally contracted to Norton) was tried and accepted with enthusiasm for its red wine. In Missouri several men of more than ordinary abilities, George Husmann, Hermann Jaeger, and Jacob Rommel, particularly, began to experiment with seedlings in which the native blood was provided by the species *Vitis riparia* instead of *Vitis labrusca*. Out of this work came the Elvira (still grown to some extent for white wine), Montefiore, and many others.

And about this time, too, the grape called Delaware came to public notice, was acclaimed as a superior grape for white wine, and has ever since shared honors with the Catawba as being the best of the "standard" American white-wine grapes.

This fact is to be noted about all of the varieties so far

[1] The Isabella, though its fruit is now deemed inferior, has remarkable adaptability; and was even put to some use by Dr. William Popenoe, the specialist in subtropical horticulture, for the breeding of a new race of grapes adapted to the Caribbean countries.

mentioned: they were wild seedlings, either pure varieties of native species or accidental hybrids involving two or more species, which were discovered by keen-eyed men and brought from nature into cultivation.

It was in 1843 that the possibility of *artificial* grape-breeding was first demonstrated, by the introduction of a grape called Diana. This was a seedling of the Catawba, the result of a deliberate effort to improve on the Catawba itself. From then on great strides were made in actual breeding. John Fiske Allen introduced a cross between the Isabella and the *vinifera* table grape Golden Chasselas in 1854. E. S. Rogers, a hybridizer of Roxbury, Massachusetts, introduced a whole series of such hybrids, still known collectively as the Rogers Hybrids. Rommel in Missouri, and Hermann Jaeger, worked along this new line with their *riparia* material. Jacob Moore, of New York, introduced a number of his own hybrids, some of which are still grown in a limited way as table grapes.[1]

In the Southeast likewise other hybrids were introduced, of which the Herbemont for white wine and Lenoir for red wine were by far the most popular.

This story of the American grape-breeders could be much expanded. But the lines of its development have been indicated. The work continues; and indeed during the years since the repeal of Prohibition it has increased substantially, both among private investigators and among the workers of some of the State Agricultural Experiment Stations.[2] But in

[1] Moore's Early and Diamond particularly.

[2] One must exclude the Federal experiment stations. Shortly after Repeal a modest program of wine-grape investigations was begun, in cooperation with the State stations. The equipment for a model experimental winery was even bought for installation at the huge Beltsville agricultural station near Washington, D. C. Then certain influential Congressmen heard of the program and put a stop to it on threat of holding up the entire appropriation of the Department of Agriculture. The program for testing wine grapes came to a dead stop and has been immobile ever since. The fine new equipment stood for many years in its original packing cases in a Beltsville basement; and members of the Department look furtively around before they so much as mention wine grapes. Except by inadvertence, wine grapes are always called "juice" grapes in U.S.D.A. publications. Thus, though the Federal Government is presumed to inter-

drawing this part of the story to a close it is enough to mention two additional sources of these American hybrids.

One, and probably the most significant of all in terms of wine-growing, was the work of Thomas Volney Munson, a curiously ascetic man who, at a very early age, was impressed with the great possibilities of grape-breeding in this country. After several false starts he moved, in 1876, from the Middle West to the relatively untouched region around the town of Denison, Texas. Here, in the woods near by, along the ravines of the uplands and in the river bottoms, were vines of species unfamiliar to him. Along the high bank of the Red River – indeed, in whatever direction he struck out – he found exciting wild material ready for his hand. "I had found my great paradise!" he said. "Surely now, this is the place for experimentation with grapes! . . . Thus was rekindled my passion for experimental work with grapes." Munson tells his story in a remarkable book: [1]

> The pioneer originator must travel much in the woods of every section where wild grapes grow, and study the habits; search out and collect together the best varieties from every region and breed up their good properties. . . . At various times during the past thirty years, the writer traveled through forty of the states and territories of the Union, never neglecting any opportunity to hunt and study the wild plants, especially the grapes and other wild fruits. . . . Thousands of vines of nearly every species of American grape were studied growing in their native habitats.

This ceaseless process of selection from natural material provided the basis for a remarkable collection, in which nearly every American species was represented. With the help of this collection, Munson undertook two lines of work: first, the fundamental botanical labor of description and classification; and second, the breeding of new varie-

est itself in crop improvement generally, it completely disregards the immense potential source of income which a market for better wine grapes would make available for the American farmer.

[1] T. V. Munson: *Foundations of American Grape Culture* (Denison, Texas, 1909).

ties. His botanical work brought him into correspondence with botanists the world over including all the leading French ampelographers of the time, and earned him many honors. His breeding work brought him an even wider recognition. For, using the species of the Southwest, he introduced new "blood strains" of the very highest importance. Prior to that time, American grapes used in breeding had been very largely confined to *V. labrusca* and *V. riparia*. The species with which Munson worked not only broadened the range of adaptability in American hybrids[1] but introduced fruit qualities of particular value for wine-making. A few of his varieties, several of them of really big promise for the future of American grape-growing, are briefly described in Chapter XIII.

Munson's most fruitful labors were undertaken in the late 19th and early 20th century. Succeeding him as the standard-bearer for grape improvement was an institution rather than a man – the New York State Agricultural Experiment Station at Geneva, New York. Under the able leadership of Dr. U. P. Hedrick, for many years its director and now retired, this institution undertook a program of fundamental research and breeding. The research took the form, first, of bringing together a varietal collection. On the basis of this, a comprehensive descriptive book on American grape varieties was published,[2] a book that is still standard as far as it goes. The available material having been collected (though perhaps with insufficient attention to Munson's productions), a breeding program was begun. Out of this program, which has been supported with varying degrees of enthusiasm up to the present, has come a succession of varieties, some of them pronounced improvements over the standard

[1] Munson's career is a striking illustration of both the strength and the weakness of private as opposed to institutional research. No institution could have approached this work with the enthusiasm and imagination that he displayed. Yet already many of his innovations are lost to posterity, and his irreplaceable varietal vineyards are long since abandoned and forgotten.

[2] U. P. Hedrick, *op. cit.*

table sorts, some of them showing promise for wine. The emphasis on the breeding of grapes specifically for wine was somewhat increased after repeal of Prohibition, and new varieties of still more value for the wine-maker may be confidently expected in the years ahead—especially white-wine varieties adapted to short-season areas. A few of the New York Station varieties are described in Chapter XIII, though endorsement for most of them must be tentative pending a more thorough investigation of their value for wine.

2. THE FRENCH HYBRIDS

In the course of three centuries, America's richness in native grape species has therefore yielded results, of rather limited value to be sure, for the would-be wine-maker. The growing of grapes for wine is largely concentrated in California, where the *vinifera* varieties do so well. Elsewhere, the limited development of wine-growing got its start on the Catawba, the Delaware, and a very small number of other American kinds – most of which, though capable of producing tolerable wines, fall rather far short of the ideal. But within the last decade or so, the prospects for a successful wine-growing industry outside of California have been increased immensely by the introduction of a whole new category of grape varieties – the so-called French hybrids.

Since these grapes are so full of promise, and as yet so little known, it is necessary to sketch their background in some detail.

As we know, the grapes of Europe do not thrive in continental North America, partly because they are susceptible to our rigorous winters and partly because they are susceptible to an array of vine diseases. Of these diseases, one – of which the agent is an insect called the phylloxera – attacks the soft and fleshy roots of the *vinifera* varieties. Four others, commonly known as downy mildew, powdery mildew, blackrot, and anthracnose, attack the leaves, the vine, and the fruit.

Europe and Asia, the home of the *vinifera* varieties, were once free of these diseases. But during the 19th century, presumably traveling on botanical specimens, they finally reached the grape-growing areas of Europe. The resulting destruction was on a vast and tragic scale. In France, for example, a large fraction of the population earns its livelihood in normal times through some connection with the wine industry. The phylloxera swept through the great vineyard regions of France like a wave of flame, destroying thousands upon thousands of flourishing vineyards, spreading economic ruin, and drastically reducing the wine production of that wine-drinking nation. The French worked frantically to defend and replace their vineyards, but no method of subduing the destructive insect phylloxera proved effective. The solution finally emerged out of the observation that, though phylloxera is almost everywhere present in the United States, our wild native vines are not seriously affected by it. Investigation revealed that, by contrast with the tender *vinifera* roots, the roots of our American species are relatively tough and resistant to the depredations of the insect.[1] If somehow the toughness of the American roots could be combined with the qualities of fruit of the European vines (so the French reasoned) then the phylloxera problem could be resolved. But how to achieve this?

The obvious solution was by means of grafting — a practice so ancient that the directions offered by Marcus Porcius Cato in his little book *De Agri Cultura* [2] have hardly been improved upon to this day. Cuttings of our native vines were imported into Europe in immense quantities, rooted, and then set out in the phylloxera-infested lands to be used as rootstocks for the scions of the classic European varieties. The vineyards of Europe were on their way to being rescued from the menace of the phylloxera. But in this first

[1] The degree of resistance is highly variable among the American species, in some the resistance being practically perfect, and in some the resistance being only slightly better than that of *V. vinifera*.

[2] Earliest of the Roman writings on agriculture, and indeed the earliest connected work in Latin that is now extant, written in the second century B.C. and still very good and instructive reading.

wave of reconstitution endless troubles were encountered. In many cases the grafted vines failed to prosper, and after a brief life withered away. In many cases, production was far less than it had been when the vines grew on their own roots. In many cases the American rootstocks, as well as the scions grafted upon them, failed to survive. And in many cases the quality of the wine seemed inferior. The result was a widespread suspicion of the new practice. Growers, having accepted grafting as the solution of all their troubles, turned against it; and a bitter controversy, among scientists as well as practical growers, grew up about the merits and defects of grafting.

That controversy is now over, because the theory of grafting is much better understood. The sources of the early troubles were identified one by one and eliminated. In the beginning, the American cutting wood was of many species, and the wood of these species was used indiscriminately and before its actual suitability for grafting, and for growth under European conditions, had been investigated. The wood of species that are intolerant of lime was planted, with disastrous results, in vineyard areas where the soil had a high lime content. We know now that there is a great variation among the species in their ability to accept grafts and make a solid union; obviously, rootstocks that accept grafts only with difficulty should never have been used. We know also that wood from a relatively weak-growing species should never be used to provide roots for a strong-growing scion, that shallow-rooted species should never be grown in arid soils, where deep-striking roots are necessary if the vine is not to go thirsty – and so on. And we know now, too, that certain fancied consequences of grafting need never be feared. The notion, for example, that the "wild" or foxy flavors of some of our American grapes could be transmitted through the rootstock into the vine grafted upon it is an illusion. Thus, in the course of half a century, the French brought order into their grafting technique, and largely reconstituted their vineyards on a rational basis.[1] The fan-

[1] The actual technique of grafting is discussed in Chapter XI.

cied menace of grafting to the quality of France's classic wines is now recognized as a baseless fear.

In the first period of disillusionment about grafting, however, the French in their desperation had resort to another solution. This was the planting of the American cultivated varieties, in the hope that they would produce, under French conditions, wines that were drinkable even though they might be inferior to the wines that France had enjoyed prior to the phylloxera. Thus the old American varieties – Isabella, Catawba, Clinton, Othello, Noah, Delaware – were imported in huge quantities. Alas, they inspired a universal disgust among the French. The special and strongly accentuated flavors of their wines shocked and displeased the French taste. And those American varieties carrying the so-called "foxy" flavor of *Vitis labrusca* earned a special loathing. These early *producteurs directs*, as the French called them, are still grown in a limited way in backward areas of Europe, by those who are not fastidious as to quality. A few of them have even acquired their own French names.[1]

In the meantime grafting, though it has been rationalized, has not proved by any means the full answer to France's viticultural troubles. Grafting defends the vine with fine success against the phylloxera; but it offers no defense whatever against the American aerial diseases of the vine – the fungus parasites that attack foliage, vine, and fruit. The European grape varieties are highly susceptible to these diseases, and they are kept from doing great damage, in the years when they are especially prevalent, only by prodigious labors with the spraying pump. It is not at all unusual for a French vineyard in certain areas to be sprayed nine or ten times in the course of a season, so that the vineyard is bright blue with the blue of Bordeaux mixture from early spring until vintage time. These troubles aroused the hope that, by breeding, it might eventually be possible to develop

[1] The Clinton is known as Plant Pouzin; Lenoir is almost universally called Jacquez. And, incidentally, our word "foxy" has introduced a new word into the French language. The flavor is sometimes called *Goût de renard*, but more frequently *Goût foxé*.

varieties that would have the typical American resistance to these fungus diseases combined with the superior quality, for wine, of the European fruit. Certain enthusiasts, despite the glaring defects of the early American Direct Producers, set about the slow and arduous task of improving them by interbreeding with selected European varieties. The first of these French Direct Producers were not greatly superior, for wine, to their American parents. Yet they did display sufficient merit to encourage a continuance of the work. Gradually the work came to be concentrated in the hands of a few gifted and immensely patient hybridizers who by a combination of persistence, insight, and good luck have made vast progress.

To be sure, these French hybridizers had advantages over those Americans whose labors and modest triumphs in grape-breeding have already been recorded. To begin with, they had an immense and eager market, consisting of practically everybody in France who owned a patch of ground; for there is hardly a Frenchman who would not grow his own wine if it could be done without great difficulty and if the quality could be counted on. For the breeders this meant that a successful new hybrid – a grape variety easy to care for, steady in production, and yielding wine of tolerable quality – would instantly find its market.

Then, too, there was the incentive created by the lingering nightmare of the phylloxera. Who could be sure that grafting, however well it now seemed understood, might not develop new defects? And, in any case, grafting was a nuisance, an intermediate and unnatural step that every grower would prefer to avoid if he could.

Third, these French hybridizers knew what they wanted: *wine* grapes, not table grapes. Hence, in breeding, the European parent was always a wine grape, whereas most of the American grape-breeders, in choosing a European parent, have made the error of depending on European table-grape varieties. And the French hybridizers knew, too, what they didn't want. They didn't want any foxiness in the resulting wine. And this led them to avoid the use of the American species *Vitis labrusca* almost entirely, since this is the species responsible for the most pronounced of such flavors. Most of our American breeders, on the other hand, have been seduced into using this species, primarily because, growing wild, its fruit is larger and more attractive than that of most of our other species. The French breeders, therefore, counted mainly on those American species whose fruit was most nearly neutral in flavor.

Indeed, the whole approach of the French breeders has been in sharp contrast to that of their American brethren. In this country breeding has been random, casual, for the most part unsystematic, and without any very clear objective in mind. The objective of the French breeders, however distant it may have seemed when they began, was well defined. They sought, by interbreeding, to produce new grape varieties having the fruit qualities of the best European wine grapes and the disease resistance of the most resistant American species. The best of them laid out a systematic program, a kind of obstacle course, which only a small fraction of their new hybrid seedlings could survive.

To begin with, they chose the parent varieties with definite objectives in mind – the American parents being those that had shown outstanding qualities of health and vigor in definite areas of France, the European varieties being those whose health and vigor were relatively good and whose qualities of fruit the breeders sought to perpetuate.

Having performed the actual work of hybridization, and having insured the rooting of the seeds that carried the hybrid germ plasm, they raised the seedlings under conditions which only the most resistant could endure. The weaklings – those especially susceptible to fungus disease and those lacking in vigor – were ruthlessly eliminated at the end of the first season. Those that survived this first culling, and they might be a few hundred out of thousands, were planted out in vineyard rows. They were set in soil heavily infested with phylloxera, and were subjected, unsprayed, to the hazards of the fungus diseases. Since the fungus diseases vary in intensity from year to year, some of the hybridizers actually took the trouble to infect their vineyards with disease when it was not naturally prevalent. Several years are required before a seedling vine comes into full bearing. During these years there was further drastic elimination.

Then, as the fruit came into bearing, the crop was subjected to a rigid scrutiny. Those that proved shy bearers were eliminated without further ceremony. Also eliminated were those showing too pronounced flavors and those that failed to make sufficient sugar content, or were too acid, or showed a susceptibility of fruit to disease.

The very few that survived this grueling obstacle race were then multiplied in numbers sufficient to allow a test of their wine quality. Here again, many fell by the wayside. And of those that remained, there was the further obstacle of a test over a period of years, not only in the mother vineyard but in other localities as well, to see whether they lived up to their initial promise.

There has been nothing like this until very recently, in the breeding of American vines.

True, not all the French hybridizers were so exacting. The development of wine-grape hybrids in France has had its share of handicaps: premature enthusiasm, high-pressure salesmanship, outright dishonesty among nurserymen, who have sold large quantities of inferior vines under false names and numbers, and the urge even on the part of the better hybridizers to launch new novelties even before their earlier productions had been fully tried. Hence, with many, the very name of *hybride* acquired a bad odor. The worst of the French hybrids are very, very bad.

Contrariwise, the best of the French hybrids are very, very good. As with all other plants, they have limits of adaptation. A hybrid such as Couderc 7120, which is much grown in the Department of Hérault, is hardly worth growing in many other parts of France. The Baco No. 1, which grows with great success in the Loire Valley, is denounced as being a shy producer in other areas. There are inexplicable failures of adaptation — a susceptibility to some specific disease in one district in contrast to virtual immunity to the same disease elsewhere, or unevenness in ripening in one place in contrast to perfect ripening elsewhere. Many of these variations in adaptability must be laid to causes of which we are still ignorant.

Furthermore, hybrids by their very nature have certain limits. The Couderc 7120, already mentioned, does not produce a fine wine. It is grown, in an ordinary wine region, for the mass production of ordinary wine; and for that purpose it does very well. To expect what the French call a *vin de bouteille* (that is, a wine capable of acquiring bouquet and "finish" with bottle-age) from this grape is to expect the impossible. Nor have the breeders yet succeeded in fixing in any of the hybrids the specific qualities of fruit that characterize the noble families of fine-wine grapes. There are hybrids which in some respects resemble, let us say, the Cabernet Sauvignon of Bordeaux, and which yield very good wine. But it is *not* Cabernet wine, and there is no sense in pretending that it is. There is no exact hybrid equivalent of the Pinot Noir, of the Chardonnay, of the Riesling, of the Sémillon. For this reason none of the hybrids are grown in France as substitutes for the fine-wine grapes. The reputation of the Burgundy region is based on the Pinot Noir. It would be shortsighted indeed to substitute another variety for the Pinot Noir, even though the wine provided by the new variety was very fine wine indeed. For, however fine it was, it would still not be Burgundy: it would not have that special bouquet, that inimitable richness of flavor, which set the true Burgundies apart from all other wines and which well-informed wine-drinkers expect when they buy it.

In order to protect the honest producers of the classic Burgundy, the growing of hybrids in the legally delimited Burgundy areas is forbidden except for wine of local consumption. What is true of the Burgundy region is true of all delimited areas [1] of France — and indeed of other wine-

[1] A "delimited area," whose boundaries are fixed by French law, protects wine-growers within the area by granting them exclusive use of the *appellation d'origine* or regional name — in return for which their wines must conform to definite standards of type and quality. Thus no French

growing countries where there is a vested interest in wines of definite and clearly established character and reputation.

The point to remember is that most of the wine drunk in wine-drinking countries is not wine duly certified and labeled with an *appellation d'origine*. Most of it has humbler beginnings, being made in the home vineyard for home consumption, or in a small vineyard near some town for purely local sale, or in huge mass-production vineyards content with a standard of quality that is something less than "fine." And much of it is used in blending, with results that, on the whole, are highly advantageous to the consumer. Let us say, for example, that Bordeaux has a bad year (1925, 1927, 1956, or 1960, for example). The wine from the fine-wine vineyards is thin, acid, lacking in color. Still, it contains qualities of aroma and bouquet that are peculiar to the fine-wine grapes. This wine is not aged and bottled under its proper appellation. Instead, it goes to the big blending merchants, who combine it in the proper proportions with wines of complementary character brought in perhaps from Algeria or the Midi – heavier, more deeply colored, but without particular distinction. The resulting blend, carried out by one who knows his business, is a far better wine than any one of its constituents.

It is in these lesser roles that the hybrids have found their place in France. They are used increasingly for the mass production of ordinary wine. They are used to provide *vins de coupage*, or blending wines. They are used for the production of local wines.[1] Perhaps most important of all, they

wine may be called Champagne unless it is made of grapes grown within the delimited area of Champagne. And even though grown within the region, it must still be made according to the prescribed method from prescribed grape varieties.

[1] Anyone driving along Route Nationale N° 74 – the main road between Dijon and Marseille, which passes along the foot of the famed Côte d'Or – can see huge vineyards growing on the east side of the road as well as on the west. The vineyards on the west side of the road are peopled with the Pinot Noir and are the source of all true Burgundy. The vineyards on the east side of the road, where the Pinot does not do well, comprise less famous varieties, including many of the hybrids; and these vineyards on the east side of the road account for the wine which the people of Burgundy actually drink.

have found their place with the part-time wine-growers, the peasants and villagers who keep their own small vineyards, and make their own small vintages each year for their own use. And here and there one sees signs that certain hybrids are beginning to develop more than local reputations for themselves – that, with time and custom, their wines will be deemed sufficiently distinctive to deserve the kind of protection now accorded to the classic vines in delimited areas.

A striking illustration of this may be found in the vineyard region of the Vaud in Switzerland. The white wine of this region has an established reputation, and no wine from hybrids has a right to bear the name. With red wines the case is different. The traditional red-wine varieties grow only with difficulty, and their wine is not of superior quality. After many years of testing, the authorities have authorized a certain number of the better hybrids for the production of red wines in the region as substitutes for the traditional *vinifera* varieties.

All this may seem needlessly detailed. Yet it has a direct relation to the prospects for wine-growing in the United States. The French hybrids were bred for typical French conditions. But the fact remains that the blood of the disease-resistant, winter-hardy American species is carried in the French hybrids. Knowing them, it is reasonable to suppose that some among the good French hybrids, transplanted here, would display their American heritage to good advantage, resisting the vine diseases of the growing season and surviving our cold winters without difficulty.

Oddly enough – though perhaps not oddly since American breeding has been so random an affair – the French hybrids were all but unknown until recently in this country. A few existed in varietal collections in California, at the Geneva (New York) Experiment Station, and elsewhere. But they grew largely unobserved, their qualities unrealized by most of those who had occasion to examine them, and hence unappreciated. Only in the vineyards of a few amateur enthusiasts, who had begged cuttings here and there from the varietal collections and in some cases imported vines on their own initiative,[1] were these vines being grown for their proper purpose, which is the production of wine. Alone among the official experiment stations, that at Geneva was taking some interest, however slight,[2] in these varieties.

It turns out, in fact, that a great many of the French hybrids are well adapted to American conditions. They are at least as resistant to disease as the standard native varieties of grapes. Their hardiness to winter cold is, on the average, fully as great; and in their quality for wine, particularly among the red-wine hybrids, they stand head and shoulders above them. The wines of the best of the French hybrids are a revelation to anyone who has worked with such grapes as Concord, Ives, Clinton, Norton, and the rest – clean-flavored, admirably balanced, and as the French say, "quick to drink." As they have made steady headway in the vineyards of Europe against all but a handful of the very best *vinifera* varieties, so their future in America is full of promise. To plant them blindly would of course be foolhardy, since we cannot have a thorough knowledge of their qualities and range of adaptability until they have been well tested under a wide variety of conditions. But in some regions their value has already been demonstrated beyond doubt. There is ample reason for confidence that, with time, this range of splendid grapes will yield varieties well adapted to all but the most inhospitable parts of the continent.

1 Plant importation is an extremely slow, difficult, and hazardous enterprise, tightly wound with red tape.

2 That interest has quickened during the last few years, and the station is now studying the French hybrids closely.

Chapter IV

THE CALIFORNIA DISTRICTS

IT IS NOW clear that the material for a successful and expanding viticulture in North America consists of three great groups or families of grape varieties. These are the *vinifera* grapes, the American hybrids, and the French hybrids. The problem for the prospective grower is to determine which are most likely to provide him with the wine he wants – what wine grapes, in short, are most promising in his particular environment. Let us first examine the possibilities in California.

The Californian who considers setting out a vineyard has one immense advantage. He has the abundant experience of a century of commercial wine-growing to draw on, and is further fortified by a very considerable body of systematic knowledge of the adaptability of wine-grape varieties to local, even to neighborhood, conditions.

1. THE CLASSIC OUTLINE

In Chapter II some reference was made to the early introduction of the *vinifera* varieties. There is no need to trace here the complicated and sometimes dramatic history of the growth of commercial wine-growing. For our purposes, it is sufficient to say that, gradually, certain important distinctions between California's various natural regions came to be understood. By the second decade of this present century, thanks largely to the systematic work of Professor Frederic T. Bioletti, viticultural California had come to be classified into six distinct regions. These were known as the *North Coast*, the *South Coast*, the *Cen-*

tral Valley, the *Sacramento Valley*, the *San Joaquin Valley*, and the *Hot Desert.*

The *North Coast* consists of the valleys lying between the parallel coastal ranges north and south of San Francisco Bay, and is characterized in general by moderate temperatures and moderate rainfall. The best of the dry table wines, red and white, are all grown in the valleys and slopes of this region; and they will continue to be.

The *South Coast* region is merely the extension in the southern part of the State of these same coast valleys and ranges. Its viticulture is somewhat similar.

A glance at the map shows that the *Central Valley*, the *Sacramento Valley*, and the *San Joaquin Valley* are merely subdivisions of that great valley which constitutes the heart of California. Climatically the whole valley is one, but with local variations sufficiently marked to justify the subdivision. The *Central Valley*, so called, owes its special characteristics to the fact that it lies directly east of San Francisco Bay, so that its summer heat is tempered somewhat by the cool, fog-bearing ocean breezes that have access to it through the Golden Gate. This is the area from which a large proportion of California's table grapes come, especially the Flame Tokay. But in addition it has an enormous acreage of wine grapes, mostly of the high-producing bulk-wine varieties. It is the American counterpart of the viticultural regions of Southern France and Algeria.

Northward into the *Sacramento Valley*, the climate becomes progressively hotter in summer and cooler in winter as the moderating influence of the Bay is left behind. This region is likewise best adapted to the mass production of ordinary wines.

South of the *Central Valley*, and likewise deprived of the moderating sea breezes, is the *San Joaquin Valley*. It is very hot, with mean monthly temperatures of more than 80° in July and August; and in most parts the annual rainfall is negligible, so that irrigation is necessary. This is the greatest raisin-producing region of the world, the grapes used for this purpose being the Muscat of Alexandria and the Sultanina. It also produces immense crops of table grapes. Finally, it has many thousands of acres of wine grapes. But, owing to the heat and the length of growing season, most of the dry wine grown in this region is coarse, heavy, and ordinary in quality. The sugar content of the varieties commonly grown runs too high, and the acidity too low, for ideal balance in dry table wine. However, these characteristics make the region suitable for fortified dessert wines.

Finally, there is the *hot desert* region of Southern California. A part of this, lying east of Los Angeles in vast San Bernardino County, which includes whole deserts and mountain ranges within its borders, is not too hot for wine-growing; and there was a time during the 19th century when it dominated wine-growing in the State. Planted with appropriate hot-country varieties, it can produce good dry wines of somewhat better than average quality, though its wines always have a tendency to be flat and heavy. It is very well adapted to sweet wine production. It is in this area that the great 5,000-acre Guasti vineyard, "the world's largest vineyard," was located. The rest of the Hot Desert country is not appropriate to wine-making, though the vineyards of the Imperial and Coachella valleys do a lucrative business in shipping extra early table grapes to the Eastern markets.

2. THE NEW SYSTEM OF REGIONS

That is what might be called the classical outline of California's viticulture; and in a broad sense this outline is not likely to be disturbed. Nevertheless, in the years since the repeal of Prohibition, this scheme has been considerably refined, particularly by the work of Amerine and Winkler.[1] The point is that within the broad outlines there exists an infinity of important local variations. We have seen how the Golden Gate gap, by letting the sea breezes into the great central valley, profoundly modifies the climate of certain sections of that valley. Influences of the same general order affect smaller areas just as profoundly. For example, in the valleys of the North Coast region, those parts lying closest to the Bay are coolest, whereas mean summer temperatures rise steadily the farther one gets from the Bay. Again, elevation in the hilly and mountainous areas plays an important role, an elevation of a few hundred feet sometimes being enough to spell the difference between ideal climate and one that is too hot for high wine quality.

During some three decades Amerine and Winkler have studied these variations methodically and in a good deal of detail. Their method has been to gather samples of all available grape varieties from every principal grape-growing part of the State. These samples they have converted into small batches of wine, analyzing them chemically and sub-

[1] M. A. Amerine and A. J. Winkler: "Composition and Quality of Musts and Wines of California Grapes." *Hilgardia*, Vol. 15, No. 6. Berkeley, 1944.

jecting them to periodic tastings[1] in the course of aging. Altogether they have compiled thousands of tasting records of these wines, and have thus been able to form a clear picture of the capabilities of hundreds of grape varieties growing in various locations all through the State.

On the basis of their investigations, Amerine and Winkler have constructed a new regional system for California viticulture. But it is a system that supplements, rather than supplants, the geographical outline that has already been made. For the Amerine-Winkler system of "regions" is based on a single factor. It takes no account of rainfall, of winter temperatures, of soil. The sole basis on which they set up their five regions [2] is the summation of heat as degree-days above 50° F. for the period April to October inclusive [3] — that is, the total heat that is available to a vine during its growing season. Thus Region I comprises those areas having fewer than 2,500 degree-days — which is to say, the coolest parts of the State; Region II consists of those areas having 2,501 to 3,000 degree-days; Region III, 3,001 to 3,500 degree-days; Region IV, 3,501 to 4,000 degree-days; and Region V, those blazing areas that suffer under more than 4,000 degree-days. Amerine and Winkler offer the following typical localities where these temperature ranges may be found.

Region I: Napa and Oakville in Napa County; Hollister and San Juan Bautista in San Benito County; Woodside in San Mateo County; Mission San José in Alameda County; Saratoga in Santa Clara County; Bonny Doon and Vinehill districts in Santa Cruz County; and Guerneville, Santa Rosa, and Sonoma in Sonoma County.

Region II: Soledad in Monterey County; Rutherford,

[1] There is a point in the analysis of wines when chemistry ceases to be useful and the relatively inexact analysis by tasting is still supreme.

[2] To be referred to always by Roman numerals as Regions I, II, III, IV, V.

[3] A degree-day is calculated as follows. If, for example, the mean temperature for a period of five days was 70° F., the summation would be (70 — 50 × 5) = 100 degree-days. If the mean temperature for June was 65° F., the summation would be (65 — 50 × 30) = 450 degree-days.

St. Helena, and Spring Mountain in Napa County; Santa Barbara in Santa Barbara County; Almaden Vineyard, Evergreen, Guadalupe district, and Los Gatos in Santa Clara County: and Glen Ellen in Sonoma County.

Region III: Livermore and Pleasanton in Alameda County; Calpella, Ukiah, and Hopland in Mendocino County; Calistoga in Napa County; Alpine in San Diego County; Templeton in San Luis Obispo County; Loma Prieta in Santa Cruz County; and Alexander Valley, Asti, and Cloverdale in Sonoma County.

Region IV: Guasti in San Bernardino County; Escondido in San Diego County; Acampo, Escalon, Lockeford, Lodi, and Manteca in San Joaquin County; Cordelia in Solano County; Ceres, Hughson, and Vernalis in Stanislaus County; Ojai in Ventura County; and Davis in Yolo County.

Region V: Fresno and Sanger in Fresno County; Madera in Madera County; Arena and Livingston in Merced County; and Trocha in Tulare County.

Anyone familiar with the striking local variations that prevail in California will see from this list why it is that the new system of regions supplements rather than supplants the broad geographical divisions. For it introduces a basis for *local* distinctions. To illustrate, let us assume the existence of a mountain that is located in very hot country and that is cultivable all the way to its summit. Vineyards planted at its base would be in Region V. Those planted a few hundred feet up the slope, where it is cooler, would be in Region IV; those planted still farther up the slope would be in Region III; and so on until, far above the surrounding country, a cool area properly belonging in Region I might be found. Thus within the range of a very few miles one might find *all* of the Amerine-Winkler regions, with vineyards capable of producing a range of wines running all the way from the most alcoholic of hot-country dessert wines to the lighter, tarter, and most delicate of cool-country white wines.

This is an extreme hypothetical case. But it does illustrate the point and give the prospective wine-grower his clue — which is to determine before he plants, by consulting the weather man, his county agent, and other appropriate sources, which of the five temperature regions his vineyard site is located in. Once he knows that, he may choose with some confidence from among the grape varieties recommended for that particular region. The descriptions of grape varieties in Chapter XII all carry recommendations as to their adaptability to Regions I through V.

A definitive study of California's grape-growing possibilities awaits much more of the kind of sampling on which the Amerine-Winkler conclusions are based. For it will be found that other factors besides the summation of available heat during the growing season will have their effect on the resulting wines. For example, it matters very much what the distribution of available heat may be in the course of the growing season. Not only the total annual precipitation but the distribution of rainfall throughout the months of the year is important. There is reason to believe that winter temperatures have their effect, too — vintages following unusually cold winters tending to be better in quality than vintages following unusually warm winters. Finally, much remains to be done in California on the effect of soil variations. All this is a long way off. In the meantime, the prospective wine-grower in California, in contrast to those who would plant vineyards elsewhere in this country, is fortunate in having so sound a body of information as is already available.

3. PRESENT AND FUTURE

In addition to the results of scientific investigation, the prospective California wine-grower has as a guide the experience of practical grape-growers and commercial wine-growers. If his object is primarily to produce wine of superior quality, however, he will be wise not to follow too literally the practices and recommendations of these men, for there is much to criticize in the actual practices of California's grape-growers and wine-makers.

To begin with, a great deal of California's wine is made

from inferior or badly adapted varieties — not always through the fault of the growers themselves. Perhaps the most striking illustration of this is the ubiquity of that inferior grape, the Alicante Bouschet. This grape is, in a sense, the child of Prohibition. For though its wine is inferior it does have certain virtues that gave it value in that confused period. Chief of these was the firmness and toughness of its fruit, which allowed it to withstand shipment to distant eastern markets and arrive in good condition. In consequence it almost invariably brought higher prices from the domestic wine-makers of the time than did the better grapes. Its wine was popular too, because of its extremely deep color; and finally it is a highly productive variety. Today, however, the Alicante Bouschet is a drug on the market; and no prospective wine-grower ought to be misled by its prevalence into planting it.

Another serious defect of California viticulture is the habit of looking on some varieties as "dual purpose" or "triple threat" grapes. The grape called Thompson Seedless or Sultanina is a popular table grape; but surpluses can always be converted into seedless raisins; and, failing both of these markets, they can always be sold to the wineries for the production of a nondescript white wine. Growers find it hard to resist the temptation to plant varieties for which there is more than one outlet. Hence a large proportion of California's wine is made of varieties that are not particularly well adapted to wine-making. It can hardly be surprising that such wines are ordinary; and they are made no better by being put into a bottle labeled "Riesling" or "Chablis."

A characteristic of wine production in California that also has certain unfortunate consequences is the very general separation of the related arts of wine-making and grape-growing. There are many exceptions, to be sure — grape-growers who make their own wine and are therefore very much alive to the importance of growing the best varieties, But the large pattern is based on the separation of these two functions. This was illustrated by the invasion of the dis-

into the California wine-growing industry during World War II. Distillers buy the grains from which they produce their whisky: they would not dream of growing it. They look on themselves as "converters" rather than farmers. Hence, when they bought into the wine industry, their emphasis was on the purchase of wineries and stocks of wine. Their purchases of actual producing vineyards were trifling.[1] And in this they were only following the dominant tradition. The wineries by custom obtain their grapes from independent growers. Although this shifts the burden onto the growers in case of either surpluses or bad crops and allows the wineries to adjust their purchases to their actual requirements, it has the defect of limiting them to those grape varieties which are actually available. If the only grapes available are Alicante Bouschet or Carignane, then the only wine they can make is Alicante Bouschet or Carignane wine.

A corollary defect of this system is that the growers, having no responsibility for the finished product, have little or no interest in the kind or quality of their grapes, provided they are able to obtain big crops and good prices. It is a matter of complete indifference to them whether they are growing grapes for shipping, for raisins, or for wine—or, for that matter, whether they are growing grapes or avocados. This lack of interest in the varietal question among growers has been perpetuated by the unwillingness of the wineries until very recently to pay sufficient differential for superior wine grapes. The grower has no incentive whatever to raise shy-producing fine-wine grapes such as the Pinot Noir or the Cabernet when inferior varieties capable of producing three or four times the crop per acre bring him prices almost as good. Owing to this division of interest between the grape-growers and the wine-makers, the average quality of California wines has unquestionably suffered.

Another criticism that is well deserved by the California wine industry is the practice, by most wineries, of trying to develop a complete "line" of wines. Working with the

[1] More than one shrewd grower sold his winery but kept his vineyard.

same limited raw material (i.e., whatever grapes are available to them), the wine-makers try to produce plausible likenesses of everything in the book, from port and sherry on the one hand, right on through the whole array of conventional "types" of dry wine to champagne. Frequently they produce brandy as well. In these circumstances it must be obvious that type names become all but meaningless. No one has ever yet given a convincing definition of the distinction between "burgundy" and "claret" as these terms are used in California. Actually they stand for nothing. And this goes, too, for such commonly misused names as "chablis," "sauterne," and "Riesling." Not one bottle out of a thousand labeled "Riesling" has so much as a drop of true Riesling wine in it.

All this is by way of warning the prospective grower not to give too much weight to present commercial practices — or, rather, to make careful distinction between those commercial practices which are bad and those which are good. For, though the California wine industry has its defects, it also has its virtues; and there are many honorable exceptions to every criticism mentioned above. From one end of California to another there are grower-wine-makers whose interest is in quality and who are very sensitive indeed to the matter of varieties. Some of these operate on a small scale and are not generally known as yet; some of them [1] operate on a very substantial scale. All of them are keenly alive to the importance of discrimination among varieties. All of them are sharing in the pioneering work of finding the best possible varieties for each of California's many regions. The prospective grower will learn much and profit greatly by the advice of such wine-growers as these. Their counsel is as sound and generous as the wine they make.

[1] The vineyards and wineries of Wente Brothers in Livermore Valley, Almaden in San Benito, Martini, Beaulieu and Inglenook in Napa, and Paul Masson in Santa Cruz are favorably known by everyone who has inquired with discrimination into the matter of California's wines; and some of the wines they produce are genuinely "fine." They are all "independents" except Paul Masson, which is the subsidiary of one of the "big three" whisky firms. (Inglenook was bought by Gallo in 1964.)

One development of the past several decades has yielded important results in the improvement of California's wines. This is the emphasis, on the part of certain enlightened wine merchants, on "varietal" wines. Instead of accepting wines under such noncommittal labels as "claret" or "burgundy," they have insisted on accurate labeling with the actual name of the dominant grape from which the wine is made. Thus the wines of Cabernet and Chardonnay, of Sémillon and the other better grapes, now enjoy their special market, and one that pays premium prices. This, in turn, has stimulated a demand from the wineries for greater supplies of these grapes; and gradually this wholesome influence begins to seep back to the growers. At present there is a strong emphasis in California (and in the long run it will be sound business) on the planting of better grape varieties.[1]

But the point likely to interest newcomers to wine-growing whose chief interest is in quality wine is that the possibilities of California are still only partly explored. The small grower who studies the varietal question carefully, and who combines all relevant information from both scientific and

[1] True, the emphasis on "varietal" wines has its comic side. For, as we have seen, blended wines are frequently superior to any one of their constituent wines. Wine from an inferior grape acquires no added virtue when it is not blended but is marketed under its own name. I cite the ludicrous case of a wine bearing the pretentious name Grand Noir de la Calmette which was recently available in the New York market. The grape variety Grand Noir de la Calmette is known the world over as a blending-wine grape; no discriminating wine-maker would dream of bottling it alone. Yet that's what was done, in deference to the fad for "varietal" wines.

Some of the world's greatest wines, furthermore, are traditionally made by blending several varieties. The true Sauternes wines are perhaps the best-known case of this, consisting by tradition of 4/6 Sémillon, 1/6 Sauvignon Blanc, and 1/6 Muscadelle de Bordelais. As another illustration I may cite the classic blend used for the production of that famous wine of the Rhone Valley, Châteauneuf-du-Pape. It consists of four *groups* of grape varieties: the first, Grénache and Cinsaut in the proportion of 2/10 of the whole, to give "*la chaleur, la liqueur, et le moelleux*"; the second (4/10), Tinto-Mourvèdre, Sirah, Muscardin, and Vaccarèse, to give "*la solidité, la conserve, la couleur avec un goût droit et désaltérant*"; the third (3/10), Counoise and Picpoul, to give "*la vinosité l'agrément, la fraîcheur, et un bouquet particulier*"; and the fourth (1/10), Clairette and Bourboulenque, to give "*la finesse, le feu, et le brillant.*"

"practical" sources with an impulse to experiment, can add his bit to this body of knowledge—and in the meantime provide for himself, or for the market, wines which fall into none of the easy present categories but are peculiarly his own.

4. EXTENSIONS OF CALIFORNIA

So far in this chapter, all the discussion has had to do with California. Actually, there are certain other limited areas in which the *vinifera* grapes succeed very well. These are in the southwestern part of New Mexico, with an extension running up along the Colorado River Valley and a narrow belt running down along the Rio Grande in Arizona and Texas. These extensions of the California viticultural area have not been very generally exploited, owing chiefly perhaps to the distance from markets. However, *vinifera* grapes have been growing in southern New Mexico, Arizona, and southern Texas for centuries. Conditions of low humidity almost eliminate the disease problem, and good crops may be produced under a rational system of culture. It must be said that in general the choice of varieties grown in these regions has not been enlightened, from the standpoint of the wine-maker. Hence it is not easy to make positive recommendations. The field is wide open to the experimenter; and in choosing his varieties for trial he must concern himself chiefly with the length of growing season in his own particular area, choosing from among the varieties whichever seem the most appropriate. In areas of strongly alkaline soil, he should look to the choice of appropriate rootstocks.

Vinifera grapes have also been grown in a limited way in parts of Oregon, Washington, and Idaho. The desiderata in these States are relative lack of humidity, length of growing season, and winter protection. Only short-season varieties should be grown.

Chapter V

THE OTHER DISTRICTS

IT IS a long jump from the *vinifera*-based viticulture of California to the entirely different viticulture based on the hybrid varieties.

It is a long jump in terms of the grapes themselves. It is a still longer jump in terms of public attitude toward wine-growing. Californians have learned to think in terms of wine. It provides a livelihood for tens of thousands of people, and it is an important element in the State's agricultural and business life. But elsewhere in the United States – that is, east of the Rockies – the related occupations of growing grapes and making wine are remote from the concerns of the average person. The production of wine in California runs around 170,000,000 gallons per year. The production of wine in the rest of the United States averages barely 5,000,000 gallons a year. In California, furthermore, the huge production is fairly concentrated. Elsewhere, the relatively slight production is widely scattered. Slight as it is, the diffusion of this production makes it still less evident to the layman. Every Californian has at least seen a vineyard of wine grapes; most non-Californians would be unable to recognize one even if they saw it.

So wine-growing is not a part of the life of most Americans. Wine in most of America is a citified beverage, drunk mainly by the urban proletariat of Italian extraction and by relatively small numbers of cosmopolitans and ex-GIs, and left severely alone by the vast middle class of Americans, whose knowledge of it is rather generally restricted to an occasional glass of sherry and some champagne at wedding receptions and other ceremonial functions. True, the use of wine is increasing considerably, and the typically American device of advertising promises to broaden its use much farther. The fact remains that for most native-born Americans wine is something exotic that is bought by the glass or bottle when splurging, a luxury the use of which is not well understood and which, therefore, is all too frequently disappointing. And in rural areas the bulk of the people, far from seeing the potential value of wine-growing to American agriculture, tend to look upon wine as something obscurely dangerous if not actually a weapon of Satan.

Hence, outside the *vinifera*-growing parts of California, the prospective wine-grower can hope for little practical help from already existing wineries and vineyards. The vineyards, small and scattered for the most part, are devoted primarily to table grapes for local markets. The men who cultivate them rarely know and understand wine.[1] As for the eastern wineries, they are few and far between, and fall into several categories. On the one hand there is a little group of large and prosperous wineries specializing in sparkling wine. Their wines are for the most part very good, made according to the traditional champagne method, their basis being the old familiar varieties Catawba and Delaware, with smaller amounts of Elvira, Isabella, and a few other grapes. Their blends are the result of nearly a century of experience; and though better blends,[2] involving other varieties, can doubtless be devised, those normally employed are nevertheless adequate; and the wineries are so familiar with their characteristics that there are no longer any serious problems in the rather complicated technique of champagnization. The attitude of these wine men (and it is understandable) is to let well enough alone. They have an

[1] Even in such wine-growing districts as the Finger Lakes region of New York State or the country around Sandusky, Ohio, one rarely sees a bottle of wine or hears wine spoken of.

[2] For example, these wineries habitually bring in large quantities of cheap, low-acid California white wine to blend with their own more acid varieties. Some of the newer low-acid hybrids could provide wine as suitable for this purpose and probably better in quality.

appreciative market for their wine, and their chief concern is less with finding better grape varieties than with obtaining bigger crops of the grapes they are familiar with.

In addition, there are a number of wineries specializing in still wines from the old familiar American varieties. Some of their white wines are good. Their red wines are uniformly poor. With one or two exceptions, these wineries show little curiosity about the possibilities of the newer grape varieties. So far as their white wines are concerned, there is a rational explanation for this: though these could be better, they are nevertheless potable and readily marketable. It is less easy to understand their lack of interest in the possibilities of better red-wine grapes. The red wines that they sell (under the old delusive names of claret and burgundy) serve chiefly to drag down the reputation of *all* American wines, being bought for the most part by naïve and unwary customers who are merely looking for "a bottle of wine" and are unable to anticipate the ugly surprise in store for them when they open the bottle. In a land of 180,000,000, a good many such bottles of wine can be sold each year to nonrepeating customers—thus supporting the wine-maker in his mistaken assumption that he actually has a market for his product.

Finally, here and there, but chiefly in New York, Michigan, Ohio, the deep South, and the Ozarks, there are a good many hole-in-the-wall wineries devoted to the manufacture of coarse sweet wine with a powerful "kick" – a kind of rural substitute for whiskey. Any grapes will do for this purpose – Concords, Scuppernongs, and so on. Indeed, grapes are hardly necessary; any fruit, when loaded down with an excess of cane sugar and fortifying alcohol, will do as well.

Thus it is obvious that the prospective wine-grower, bent on making light table wines of good quality, can expect little guidance from the experience of the commercial wineries of the East. Likewise, he can look for little help of a scientific character. Our State Experiment Stations, with one or two exceptions, know nothing of wine grapes and wine-making and care less.

For these reasons, the prospective wine-grower east of the Rockies is much more the experimentalist than his California brother. Nine times out of ten, he is breaking new ground, pioneering a new crop, when he plants his vineyard – preparing to make wine in a neighborhood where none whatever is made at present. This has its disadvantages: he cannot plant a given grape with the calm assurance that, well tended, it will yield wine of a predictable type. Yet the element of adventure, the factor of the unknown, has its special charm; the chance of utter failure is unlikely if fundamental limitations are observed; and success for a pioneer is especially sweet.

Yet the very fact that there are so few precedents requires that the grower make sure of his fundamentals. To avoid gross pitfalls, he must be clear on those basic conditions of soil and climate without which no grapevines can grow and bear fruit successfully.

2

Climate. Those elements of climate which matter most to the grape-grower are temperature, precipitation (rain or snow), and sunshine.

As for temperature, winter temperatures as well as summer temperatures are important. Grape varieties vary a good deal in their winter hardiness, the *vinifera* varieties being killed to the ground by temperatures ranging much below zero. The hardiness of the hybrids depends, in a general way, on the amount of *vinifera* blood in their makeup, though this is not necessarily a universal rule.[1] Cold, drying winds can be more injurious to grapevines than still cold. Sudden cold snaps, violent fluctuations in temperature, can be more injurious than steady cold.

As for summer temperatures, the regions east of the Rockies have not been analyzed sufficiently from the wine-grower's point of view to provide any precise recommendations

[1] Certain hybrids with a very high proportion of *vinifera* blood have proved surprisingly winter-hardy, thus leading to the theory (as yet unproved) that hardiness may be a specific heritable factor.

in terms of degree-days, as in the case of California. In general, regions whose growing season is less than 110 days (mostly mountainous country and some areas along the northern fringe of the United States) cannot mature grapes successfully. Some of the early-ripening varieties may be expected to mature sufficiently for wine in a 110-day growing season. But to insure well-ripened and regular crops, even with early-maturing varieties, an average growing season length of 140 days is desirable. In areas enjoying longer growing seasons, it is better not to plant the earliest-ripening varieties; for good wine quality is best assured when the final period of ripening takes place during progressively cooler weather. Grapes ripening in high heat ripen badly; and their vines are not so long lived. In Chapter XIII, in which the various hybrids are described, each variety is identified as early, early midseason, midseason, late midseason, or late. Roughly speaking, these varieties will ripen under average growing seasons of 130 days, 150 days, 170 days, 180 days, and 200 days or more, respectively. Yet this cannot be taken as an absolutely safe guide, since the distribution of temperature within the growing season is important, as well as the summation of heat during the growing season. To illustrate, many parts of British Columbia enjoy a long growing season, but it is so cool that the grapes lack sufficient total heat to ripen properly.

One factor that greatly affects the prospect of successful grape-growing is the presence of special thermal conditions occasioned by large bodies of water and mountain slopes. Large bodies of water, by maintaining cool temperatures in the spring, inhibit growth until the danger of frost is past, and by their warming influence in the fall may delay considerably the arrival of killing frosts. In some mountainous areas (there is a striking illustration of this in the region around Tryon, North Carolina) conditions of air drainage have somewhat similar effects. The cool air, sliding down hilly slopes into the valleys below, forces the warm air upward, with the result that the slopes escape frost and the growing season is thus considerably lengthened. Owing to such conditions, areas ideally adapted to grape-growing are sometimes found in the midst of areas having much less promise.

Another climatic influence, but one about which little is known, is the effect of latitude on grapes of various species parentage. In the higher (that is, more northerly) latitudes, growing seasons tend to be short, but days during the growing season are long. In the lower latitudes [1] the growing season is longer, but the actual hours of daylight at the height of the growing season are fewer. Just how this affects the ripening of grapes is not yet clear, but quite probably it is an influence in determining the adaptability of grape varieties. Grapes containing *labrusca* or *riparia* parentage (species most at home in the higher latitudes) rarely succeed in the lower latitudes.

As to rainfall, for good health and productivity grapevines require a yearly minimum of 20 to 30 inches (including the equivalent in snow or irrigation). But the distribution throughout the year is important too. Ideally, the greater proportion of the rainfall takes place during the dormant season, as it does in California. But in distribution there is tremendous variability as between different parts of the United States, not only as regards any given month but as between the same month in one year and in another. An excess of rainfall during the period of blossoming may adversely affect the setting of a crop. An excess of rainfall during the summer months may cause trouble not only by promoting vine diseases but by washing off the sprays with which disease is combated. An excess of rainfall during the ripening period may cause rotting of the fruit (in the case of susceptible varieties), and may upset the balance of the fruit's composition by preventing the attainment of its normal sugar content. In areas of high summer precipitation it is therefore important to plant varieties notable for their good health, regularity of bearing, and resistance to disease. In areas subject to drought, the grower, unless he has recourse to irrigation, plants varieties that have demonstrated their ability to

[1] That is, as one approaches the Equator.

get along on a minimum of moisture, and also takes care to spread his vines farther apart, in order to allow for the development of extensive noncompeting root systems. This device is ordinarily practiced with success in the dry Southwest.

Humidity, as distinct from rainfall, also conditions grape-growing, providing the ideal medium for the development of the principal fungus diseases. Hence only varieties of good resistance should be planted in humid locations. Humidity also has a special effect on the productivity of certain varieties. The grape Seibel 1000 is a case in point; for in humid locations it has a tendency to drop its berries, sometimes in quantities sufficient to reduce the size of the crop considerably. In the case of another French hybrid, Seibel 6905, humid conditions promote the development of corky patches on the berries – a condition which in no way affects the quality of the fruit for wine but which gives it an unattractive mottled appearance.

These are the general considerations regarding climate. The prospective grower will study them and choose his grape varieties accordingly. But in addition he should not overlook the very great variations to be found locally. In choosing the actual spot for his vineyard he should try to avoid frost pockets, areas lacking proper air drainage, shade, and the competition of other growth.[1]

Soil. Climate and soil are usually linked, and properly, for soil is in a very real sense the product of climate. In the United States, for example, our great natural soil regions correspond very closely with the belts of temperature and rainfall. Abundant rain and much heat have robbed the soil of the Southeastern States of much of its fertility. The great North-South prairie belt, with its small annual rainfall, has characteristics that are wholly distinct. Yet in spite of these resemblances there exist in every major soil region an infinite number of variations in the chemical and physical qualities of the soil, depending upon the nature of the origi-

[1] A large tree can sharply affect the fertility of a surprisingly large area of surrounding land.

nal geological structure and the degree of the soil's geological "maturity."

The vine succeeds on many different types of soil — is, in fact, one of the most adaptable of plants in this respect. Nevertheless, for the best results, some attention must be paid to its physical properties. A very light soil is easy to work and provides good drainage, but it may be subject to drought and in cold climates may sometimes expose vine roots to danger of freezing. A heavy, compact soil is "cold" in the spring and hard to cultivate, bakes badly during the hot months, and seldom provides good drainage — a physical characteristic that most grape varieties prefer. Vines dislike wet feet, and, mainly for this reason, they like gravelly or rocky soils; a rocky soil also has another characteristic that is especially valuable in the cooler regions: it retains much of the sun's heat.

As for soils that are rich in humus, the French will have none of them. Humus, which is decayed organic matter, has great virtues, and soils rich in humus are very fertile and yield abundant crops. But in wine-making, abundance is rarely the companion of the highest quality. That is why, in many great wine-growing districts of the world, the finest wines are grown on sparse hillsides, or, if the topography is not very hilly, are so frequently found on land that the average farmer would probably pass by as unprofitable.

All this should not be taken to mean that the grape-grower must steer clear of reasonably fertile land. What it means is merely that too much fertility is fully as bad for wine-grapes as too little.

Some misconceptions regarding the relation of soil types to wine quality are discussed in Chapter IX.

3

The regions for a viticulture based on the hybrids cannot, then, be determined on a basis of experience, but only tentatively on a basis of fundamental characteristics chiefly of climate. Any system of districts — for example, the one

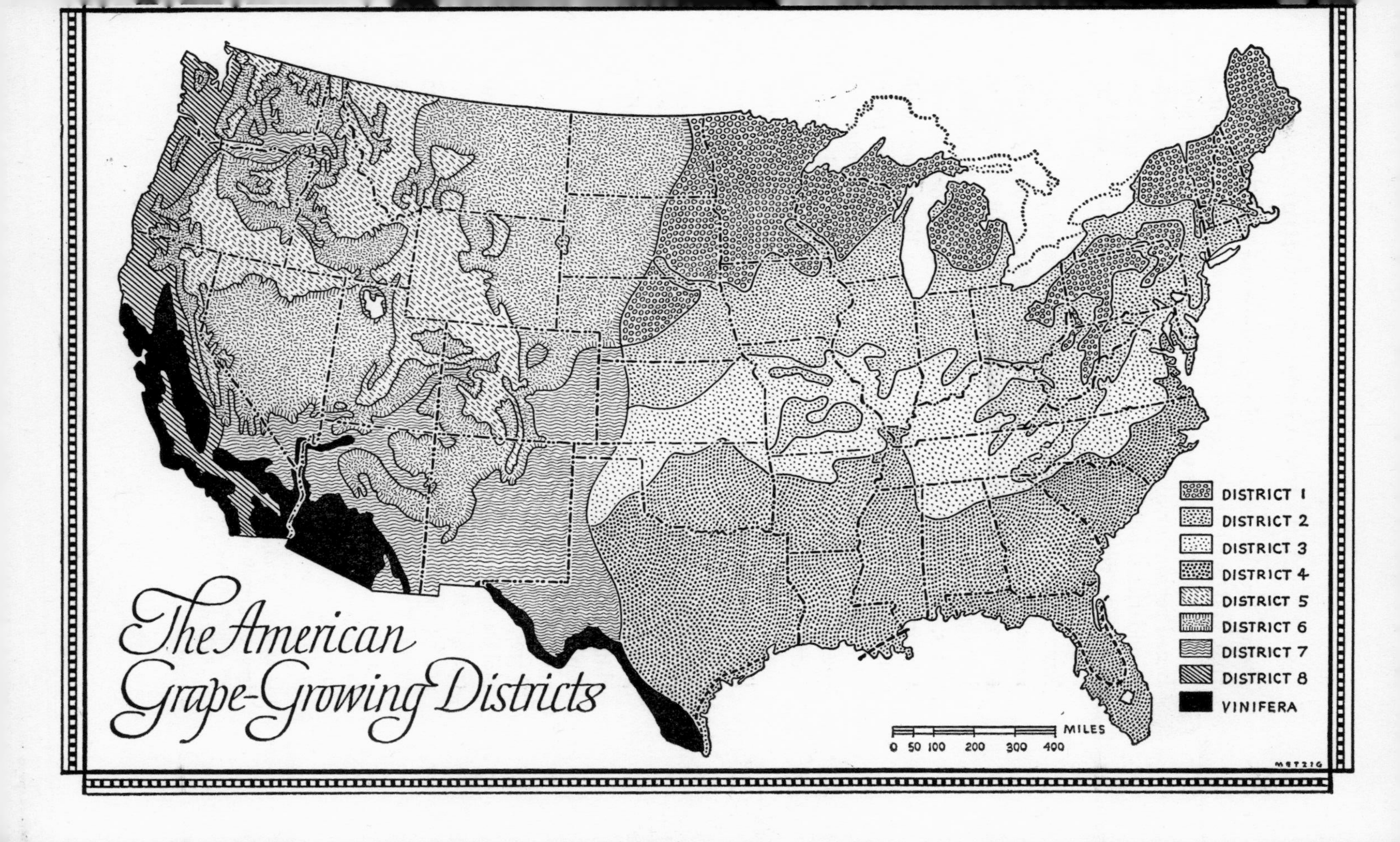
The American Grape-Growing Districts
DISTRICT 1
DISTRICT 2
DISTRICT 3
DISTRICT 4
DISTRICT 5
DISTRICT 6
DISTRICT 7
DISTRICT 8
VINIFERA
MILES
0 50 100 200 300 400
METZIG

shown on the accompanying map – must be highly tentative, and must be based on such broad grounds that it is subject to numerous exceptions and qualifications. All such broad regions or districts shade into each other, so that the boundaries should not be taken too literally. And they are subject to revision by experience. The outline of districts shown on the map [1] is only one of several such schemes that have been constructed in the past and is hardly to be taken as the last word. It cannot be emphasized too much that this scheme of districts [2] must not be taken literally, either as to its boundaries, the details of the accompanying descriptive notes, or the recommendations of varieties for planting. County agents can sometimes supplement these notes with limited advice on grapes in general, but are rarely trustworthy on specifically wine-grape questions. Weather stations can provide appropriate local qualifications and amendments.

District 1. We begin with a note on a district that is sharply limited in its grape-growing possibilities by two factors: shortness of growing season and low winter temperatures. It comprises upper New England, much of the northern half of the Appalachian mountain system, all of northern Michigan and much of southern Michigan, and a substantial chunk of the northern Great Plains area. The growing season ranges from 90 to 150 days, and winter temperatures run lower than in any other natural area of the United States. Rainfall is everywhere ample, ranging from 25 to 45 inches; and in most parts humidity is fairly high, requiring consistent disease-control measures. The grapes best adapted are hardy *riparia* hybrids, such as the grapes of the Alpha group,[3] and among the French hybrids only those with a

[1] This map is adapted from C. A. Magoon and Elmer Snyder: "Grapes for Different Regions." U.S.D.A. Farmers' Bulletin No. 1936, 1943. This same bulletin has likewise been drawn on for some of the facts contained in the notes on districts.

[2] These general districts will always be referred to, in Chapter XIII as well as this one, by arabic numerals – to assist in distinguishing them from the system of California regions, the latter being always referred to by roman numerals.

[3] Hardly worth bothering with, actually.

substantial portion of *riparia*, and possibly *rupestris*, blood should be tried—such as Joffre, Foch and Seibel 13053. Even so, in the most intemperate parts special measures for winter protection[1] may be necessary.

District 2. This district contains all of the principal present grape-growing sections outside of California. The map, though necessarily arbitrary, gives a pretty good indication of its extent. In general it may be said to include the more clement parts of New England, the southern Appalachians, large areas in the North Central States, and an extension west of the Mississippi which includes most of Iowa, the northern edge of Missouri, parts of Kansas and Nebraska, and the Ozarks area. The areas where grapes are grown most extensively are the Chautauqua Belt extending along the southern shore of Lake Erie, the Finger Lakes district of central New York, the region around the western end of Lake Erie, the region along the southeast shore of Lake Michigan centering in Paw Paw, Michigan, the Ozark grape-growing sections, and scattered grape-growing districts along the Missouri River well into Nebraska. The length of growing season from frost to frost runs from 150 to 180 days on the average, and in all but the most western parts there is ample rainfall. Most of the region has fairly high average humidity so that protection against the fungus diseases and insects is necessary; but there is great local variability in this regard. In the Chautauqua region, for example, where the breezes never cease to blow, the vines grow in superb health with relatively little need for spraying.

A large assortment of grape varieties, both the American hybrids and the French hybrids, find conditions congenial within this region. For those who enjoy the "American" flavor there are Catawba, Delaware, Brocton, Dunkirk, Elvira, Canada Muscat, Niagara, Diamond, Iona, Dutchess, Buffalo, Steuben. Of the French hybrids those grouped as early and early-midseason have proved themselves and are worth trial on a broad scale. These include for white wine Seibel Nos. 5279, 9110, 10868, and 13047, Seyve-

[1] See pp. 119–21.

Villard 5276, 12375, White Baco; for red wine Baco No. 1, Burdin 7705, Foch, J. S. 26–205, Léon Millot, Seibel Nos. 7053, 8357, 10878, 13053, Seyve-Villard 5247, and others.

District 3. Like District 2, this one is highly irregular in outline, and so variable within its limits that, with time and experience, it will unquestionably be cut up into several regions with substantially different characteristics. Hence, recommendations must be accepted with caution. It includes, on the north, the almost oceanic conditions of Cape Cod and Long Island and extends in a southeasterly direction to include the tidewater areas of New Jersey, Delaware, and Maryland; and from there southwest to take in the whole Piedmont area; then westward in a broad belt extending as far as the 20-inch-rainfall line in Kansas, Oklahoma, and northern Texas. Its growing season averages from 180 to 200 days, and rainfall is everywhere ample except in the most westerly section. In general, it is a region of high humidity and warm, muggy summers, hence the prevalence of fungus disease and insect pests. In the southern parts of this region many of the hybrids containing *labrusca* blood fail to do well; and all varieties showing high susceptibility to disease ought to be avoided. Yet the range of varieties worth trial is even broader than that in District 2, since the season is long enough to ripen those varieties classed as midseason, yet in most parts not too long to exclude the early-midseason varieties. Of the standard varieties being grown, Delaware and Catawba are successful in most places, and also Cynthiana, Elvira, Ellen Scott, Delicatessen and (where humidity is not too high) Lenoir. In general, the new New York State hybrids are not very successful. Many of the Munson hybrids in addition to Ellen Scott grow well, especially in the western half of the region. Of the French hybrids, those that have already proved their worth, show good promise, or deserve trial are, for white wine, Seibel 4986, 9110, 11803, Seyve-Villard 5276, 12309, 12375, 23410, Vidal 256, Meynieu 6, Ravat 51; for red Seibel 5455, 7053, 10878, 13666, Landot 244, 4511, S.V. 18283, 18315.

District 4. This district is much more homogeneous and

consists, roughly speaking, of the southeastern part of the United States extending from the tidewater land around the lower end of Chesapeake Bay almost to the Rio Grande valley in Texas. Where our Southeastern States dip into the subtropical belt (as indicated by the broken lines) most deciduous fruits, including the American grape hybrids, will not grow successfully, since they turn evergreen. This is a region of long growing season, ranging from 200 to 240 days. In most parts the rainfall is high, ranging from 45 to 60 inches. The sum of temperature during the growing season is also high, and is combined with high humidity over most of the region. Thus disease conditions are more redoubtable than in any of the other regions so far mentioned, and are complicated by much trouble with root rot and root destruction by nematodes. Unfortunately, little work has been done in determining suitable varieties.[1] Of the standard hybrids, however, a number may be grown when they are grafted on appropriate rootstocks.[2] Those standard hybrids that do succeed fairly well on their own roots are Cloeta, Delaware, Ellen Scott, Herbemont, Lenoir. But it ought to be added that in this district the early-ripening varieties are not very desirable for wine-making, even when they ripen their crops successfully, since the ripening takes place in the heat of midseason and thus complicates the wine-making. Of the French hybrids, there exists a large group that have proved themselves in those parts of viticultural Europe where the season is very long, notably in the humid area of southwestern France along the Bay of Biscay and in the Mediterranean coastal districts. Further, the parentage of many of these hybrids contains blood of several species that are found, in their wild state, in the western part of the district under discussion. Some of them should eventually

[1] The grapes native to much of this district are those of the species *V. rotundifolia*, which differs sharply in vine and fruit from the various "bunch-grape" species. They have such limited value for wine that I have thought it best to exclude the species and its many varieties from discussion in this book. The best-known variety is Scuppernong, and there are many improved varieties.

[2] See pp. 166–72 and pp. 195–97.

prove themselves. A number are already being tried in scattered locations, with promise of success. These include: Couderc 13, 7120, 29935, Seibel 7053, 8357, 13666, Ravat 51, the Galibert hybrids, Seyve-Villard 13209, 12375, 23410, 23–657, 34–211 and Vidal 256.

District 5. This is a kind of grab-bag, where one finds much of the highest and dryest land of the continent, and where, owing to altitude and low average temperatures, the season between frosts is very short, in many parts less than 90 days. It is therefore in general highly unfavorable to grape-growing. However, it contains a number of fertile, sheltered, and attractive valleys, enjoying the benefit of more heat combined sometimes with very low rainfall. Hardy, early-ripening varieties may be tried in some of these locations with hope of success, though it may be necessary to use winter protection.

District 6. This district comprises much of the northern part of the Great Plains area, plus extensions to the south and west into upland New Mexico and Arizona, and considerable areas in the intermontane region between the Sierras and the eastern ranges of the Rockies – all of it characterized by an average rainfall of 20 inches or less and average growing seasons ranging from 90 to 150 days, plus extremely low humidity. In this district, too, the character of the soil is relatively more important because a good deal of it is alkaline – and grapevines with American parentage do not, in general, succeed well when grown in alkaline soil. The most promising sites for grape-growing are in the valleys, where water is a less serious problem, and among the foothills where irrigation is available. Some success has been achieved in the arid upland parts of Arizona and New Mexico, by adopting the dry land practice of spacing the vines far apart. As to varieties, on soils which are dry but deep, those having *rupestris* blood stand the best chance of success. *Rupestris* is also more tolerant of alkalinity than some of the other species – and the species *Champini*, *Berlandieri* and *Lincecumii*, are still more tolerant. In general, only hardy, early-ripening varieties should be tried, among them those of the

Kuhlmann group: Joffre, Foch and Leon Millot where water is not a problem. Of other French hybrids Seyve-Villard 5247 and 5276 should prove adaptable on moderately alkaline or neutral soils and Seibel 5279, 7053, 8357 and 13053 and worth trial elsewhere. *Rupestris* and *Berlandieri* rootstocks should be helpful in all locations.

District 7. This comprises the southern portion of the Great Plains west of the 20-inch-rainfall line, with extensions westward into New Mexico, Nevada, and a portion of southeastern Colorado, plus scattered areas elsewhere. Its grape-growing possibilities are far superior to those of Region 6, for it combines low humidity with a longer growing season – ranging from 150 to 200 days. The limited rainfall, however, requires either irrigation or dry-farming practices. Owing to the low rainfall, this region also has large areas of alkaline soils, and in these parts either lime-resistant varieties or varieties grafted on lime-resistant rootstocks should be used.

In certain relatively mild parts of this region it is also possible to grow early-ripening varieties of the *vinifera* species, though winter protection is desirable. The *vinifera* varieties so far tried have been chiefly table grapes. However, such superior varieties as the Pinot Noir, Chardonnay, White Riesling, Traminer, and Gamay most certainly deserve to be tried by experimental amateurs – and with some prospect of success.

As for the rest, Catawba, Ellen Scott, and Lenoir have all been recommended; and, in addition, Delaware, Delicatessen, Dunkirk, and Buffalo may be tried with confidence. As for the French hybrids, they are practically unknown in these parts; but those classified as early and midseason may be tried. The Experiment Station recommends Baco No. 1.

District 8. This district is characterized by a growing season ranging from 150 to 250 days, and in general may be said to be that part of the West Coast lying west of the Sierras which is not very well adapted to the *vinifera* grapes. However, the reasons for the inadaptability of the *vinifera* grapes vary substantially from one part to another. The northern half, for example, is characterized by very high rainfall, ranging from 60 to 100 inches a year, and a growing season which, though long, is also cool and therefore fails to provide the necessary summation of temperature. Farther south, and especially along the foothills of the Sierras, rainfall is very low, but high elevation imposes conditions too cool for most of the *viniferas*. Hence, in general, only the hybrids give real hope of success – and even these not everywhere. In any case, only the earliest varieties should be tried. Some of those already being grown in a limited way, and especially in the Willamette Valley and parts of southwestern Oregon, are Delaware, Isabella, Golden Muscat, and Campbell Early.[1] A number of growers are also experimenting with the early-ripening French hybrids, including Baco No. 1, Seibel 5279 and 13053, and Foch.

In the southern part of the region, in sheltered locations, early-ripening *vinifera* varieties are grown to a limited extent, and may be expected to succeed if winter protection is provided, though they are not likely ever to be grown commercially on a substantial scale.

[1] Here known as the Island Belle, and under no circumstances to be considered a superior wine-grape variety.

Chapter VI

THE VINE

IN THE previous chapters I have done what I could to sketch in the background and the prospects of wine-growing in the United States. And something was said, too,[1] about the fruit of the vine, its general character and the ways in which the fruit of one variety may differ from another. But so far nothing has been set down in a systematic way on the vine, its several parts, the way it grows, and the art of its culture.

In spite of great differences between one species of grape-vine and another, the structural elements and fundamental growth habits are shared by all and may be discussed in common. There is first the obvious division between the underground system (roots) and the above-ground system; the above-ground system may be further divided into trunk, canes, buds, flowers, fruit, tendrils, leaves. There are good practical reasons for understanding in a general way the nature and functions of these various parts.

Roots. A vigorous and healthy root system is important because the root system has the double function of supporting the vine and of taking from the ground the water and the elements necessary to its growth. When a grape seed germinates, it puts forth a single tiny taproot. The tip of this has a tough cap which protects it from injury as the root pushes downward. The part of the root that lies just behind this cap is covered with exceedingly fine root-hairs that do the actual work of absorbing the soil solution. As the root develops, it gives off branches; these, as growth proceeds, give off still other branches, until finally the main outlines of the root system have been established. Vine roots

[1] See p. 7.

may be thirty or forty feet long. The mature root system, under certain circumstances, may plunge deep, but most of it usually remains near the surface.

Roots of the various species differ a good deal in their structure. Roots of *vinifera* (the European grape) are soft, thick, and fleshy, and easily succumb to disease. The roots of most of the American species are lean, tough, and rangy, and are not greatly troubled by disease. That is why, wherever the *vinifera* grapes are grown, they tend more and more to be grafted on American rootstocks.

The roots of various species differ not only in structure but in behavior. Some roots, for example those of the American species *V. rupestris*, plunge deep; others, such as the American *V. riparia*, have a spreading habit. These differences are important culturally because a deep-rooting vine is naturally better suited to land that is subject to drought and where the water table is low, and a shallow-rooting vine is superior where the soil lies in a thin layer over impervious rock or where the subsoil may for one reason or another be injurious to the vine.

Finally, it ought to be made clear that comparatively few vine roots have their start in seeds. Most vines are not grown from seeds but are propagated vegetatively from cuttings. When a vine cutting is planted, adventitious roots develop at first in great profusion; then gradually a few dominate and the rest die off. A vine from a cutting, therefore, does not have the regular branching development from a single taproot that distinguishes a seedling vine, but usually has two or three main roots.

Trunk. The trunk is the permanent part of the vine that appears above ground. It is the skeleton of the vine, and it is likewise the transmission system by which the food is carried as crude sap from roots to leaves, there to be elaborated and made available for the nourishment of the plant. Every cultivated vine has a permanent trunk. It may be only a few inches high, as in some methods of stump-training. It may be four feet high for trellised vines, or as tall as a tree and ranging from one tree to another, as in the *arbusti* which

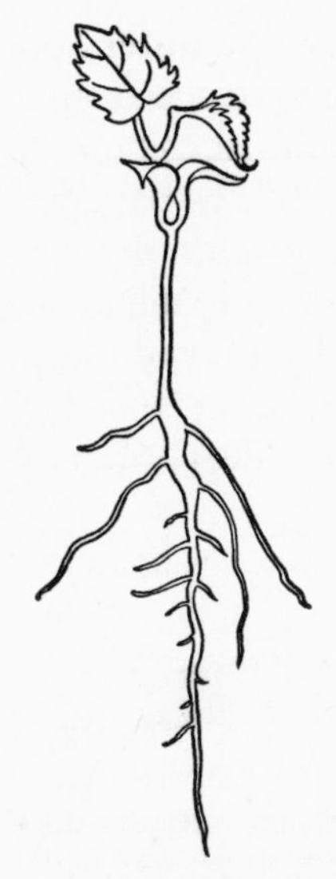

FIG. 3. *Grape seedling, showing tap root.*

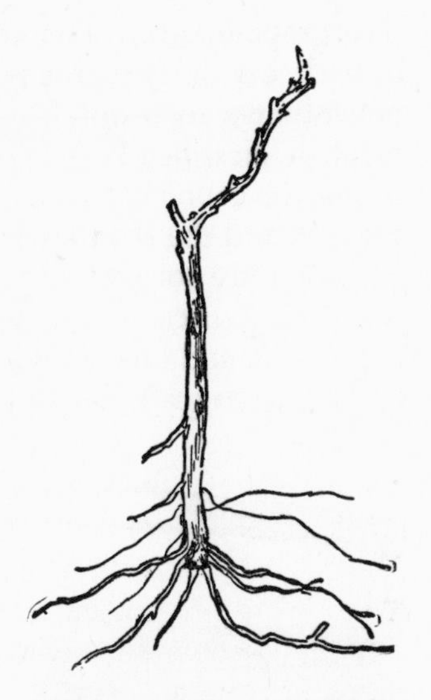

FIG. 4. *Rooted cutting, roots originating adventitiously from region of basal node.*

Cato describes and which are still to be found in some parts of Italy. And there may be more or less permanent arms or branches, which are really only extensions of the trunk. With age, the trunks frequently grow to very large girth.

Canes. The rest of the stem or woody structure of the vine – the part on which leaves, flowers, and fruit are carried – is the annual growth of long, slender canes. These develop from buds in the spring, as tender green shoots. The green shoots gradually acquire their woody character during the season's growth. On normal vines the shoots grow rapidly. The length of the mature shoot, or cane, as also its color, the character of its bark, the nature of its nodes or joints, the length of its internodes or spaces between joints, its shape in cross-section, and the appearance of the bud-eyes that occur at the nodes, all vary from species to species.

If you slice a mature cane in cross-section, enough of its structure will be revealed to satisfy the curiosity of the practical grape-grower. The cross-section shows an outer layer of bark, cut off from the rest of the cane and hence from all nourishment by a thin layer of cork. Since it lacks nourishment, it dies, becomes loose, and may presently be peeled off. In the center of the cane is the column of pith. Between the pith and the cork layer lies the cylinder of living wood, in which are located all the complex and highly differentiated structures which enable the vine to carry food and drink back and forth, to store up nourishment, to hold itself erect, and to grow in girth.

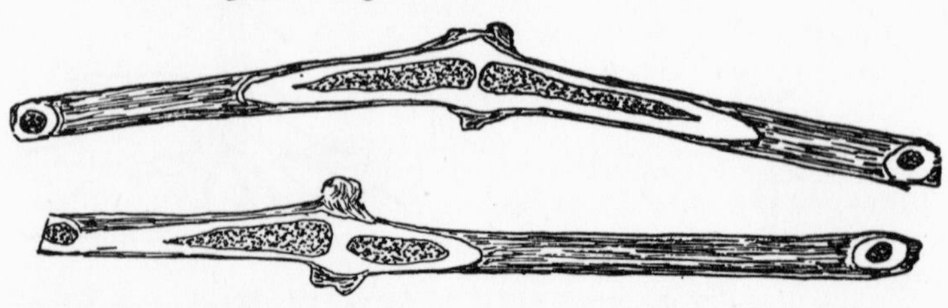

FIG. 5. *Sections of cane of two different species with sections cut away to expose cross-section at nodes.*

One thing more about the cane: it will be noticed that a woody diaphragm extends across the node. This diaphragm is present in the canes of all but two species (though it varies in thickness) and is a valuable character in the identification of species.

The pruning of the vine, which will be discussed later in detail, consists essentially in the removal of the greater part of the year's growth of cane-wood, leaving only enough to yield in the following year the crop of grapes desired.

Buds. The buds are undeveloped shoots, and are of two kinds: *bud-eyes*, which are found only on the year-old canes, and the invisible *adventitious buds* that exist in large numbers on the older wood of the vine.

Most of the adventitious buds sleep placidly throughout the life of the vine. But each spring a certain number of them awaken and give rise to young shoots or water sprouts, these tending especially to emerge from the base and head of the trunk. Water sprouts are not often fruitful and most

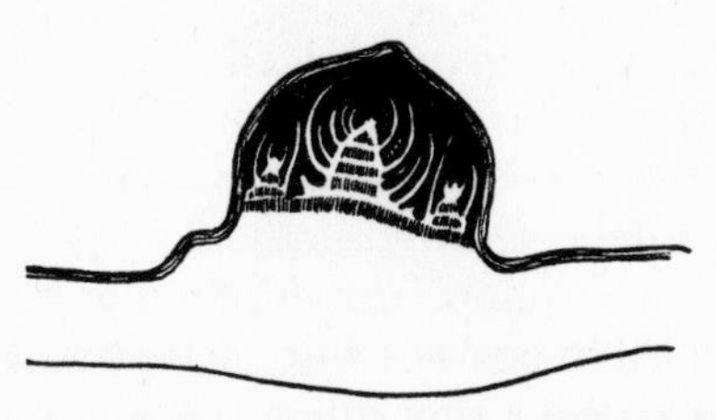

FIG. 6. *Cross-section of a bud-eye, highly stylized, showing outer scales, felty inner layer, and rudimentary shoots, leaves, etc.*

of the time they are a nuisance, robbing the fruitful canes of nourishment and giving nothing in return.

The important buds are those which occur in the bud-eyes at the nodes of canes. It is from these that the fruitful shoots develop in the spring, to bear leaves, flowers, tendrils, and fruit, finally to mature in the autumn and become the canes of the following year. These eyes, which during the winter are covered by two dark, hard scales, appear to be single buds. Actually they are compound, consisting of a central bud and two secondary buds. In the diagram above may be seen the cane in embryo: rudimentary shoots, leaves, flowers, and tendrils. The main bud develops first. The secondary buds develop later, but do not make growth comparable to that of the main bud unless the shoot from the main bud is killed by frost or comes otherwise to grief. The secondary buds of some (but not all) grape varieties are fruitful, so that even though the primary shoot is killed a crop may still be had from such varieties.

Those parts of an eye which are destined to grow as shoots are very tender; they are protected during the dormant season by a brownish cottony insulating material and are still further protected by the two hard scales that have been mentioned.

In early spring, the eyes grow noticeably plump; then with the coming of warm weather the two scales loosen, and the shoot appears. This is a lovely thing, indescribably fresh and delicate. According to the variety, the tiny young leaves are brilliant carmine, or carmine edged with gray or cream, or pale mossy green, edged with rose or violet or perhaps with green of a darker and more brilliant hue, or dull bronze. In texture also they differ, some being felty or covered with a delicate down, others glossy. The growing tip continues to develop, unfolding more leaves, then a leaf and a tendril[1] together but on opposite sides of the shoot, then a leaf and a tiny inflorescence (the flower cluster), which at this stage looks like a miniature bunch of grapes, then another leaf-and-inflorescence, and so on. Most grapes have only two inflorescences to a shoot, but some varieties regularly develop as many as four.

The shoot grows with great rapidity; the shoot itself begins to stiffen, the leaves to assume their mature form and size, the inflorescences to gain in size and become more clearly differentiated, the tendrils to grope and cling.

The inflorescence opens and is fertilized; the tiny berries appear and start to grow; by this time the shoot has concluded its first and greatest spurt of growth, and begins to have about it a look of approaching maturity. The leaves have lost that virginal freshness, are less supple, more leathery, deeper in color. The shoot now more and more resembles a cane. If one examines the axil of a leaf (where the leaf joins the shoot), it is possible to detect two small rudimentary buds. Before the summer is over, one of these will have pushed forth and developed into a *lateral* shoot, usually unfruitful but in some varieties showing inflorescences and bearing a small, late second crop. The laterals are usually short and of feeble growth; but sometimes they ripen into sound wood and survive the winter, and their buds are then as fruitful as those of a primary cane. The other developing bud to be noticed at the axil of the leaf is still there when leaves have fallen and the crop has been gathered – the bud-eye of the following year.

Flowers. The inflorescence, or flower-cluster, of the grape consists of many tiny individual blossoms, each attached by its individual stem (called the pedicel) to a larger stalk to

[1] Some varieties have a tendril opposite each leaf; others are known as discontinuous, and do not show a tendril at every node.

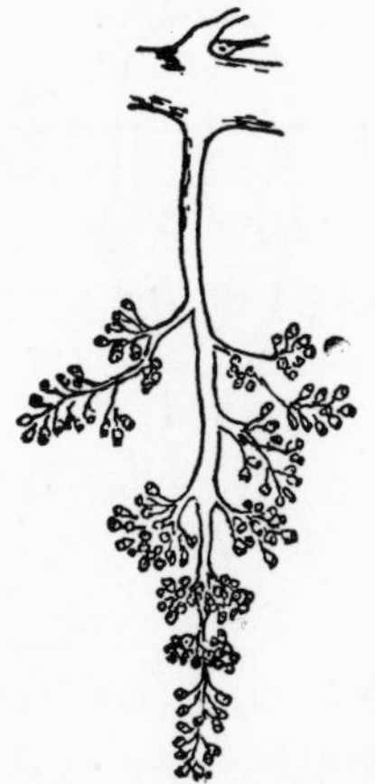

FIG. 7. *Inflorescence.*

form a compound flower. The flower clusters of the various species have their characteristic forms, which determine the shape of the bunches of ripe grapes.

Those who are not familiar with the ways of the vine sometimes confuse the first appearance of the young shoot with the blossoming of the vine. Actually, and unlike such fruits as apple and pear, the vine does not blossom until many weeks after the eyes on the canes have burst; the vine achieves most of its season's growth before it undertakes the exhausting task of producing a crop of grapes.

Most cultivated varieties have hermaphrodite flowers: the blossoms are capable of fertilizing themselves. The unopened flower looks something like a tiny green inverted punching-bag, being covered by the five little petals of its *corolla.* When the flower opens, it opens not from the top but from the bottom, the five petals loosening and separating at their base but holding tightly together at the top. The entire corolla slips off like a cap and exposes the sexual organs: the conelike female *pistil* consisting of *ovary*, *style*, and *stigma*, and the surrounding masculine *stamens* (five or six usually) consisting of *filaments* and their surmounting *anthers.* The anthers consist of tiny sacs of pollen. The *nectaries*, which give off the perfume, are small protuberances located at the base of the pedicel, between the filaments.

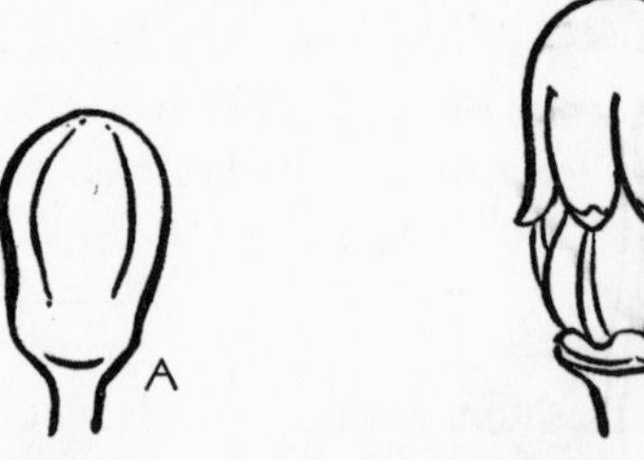

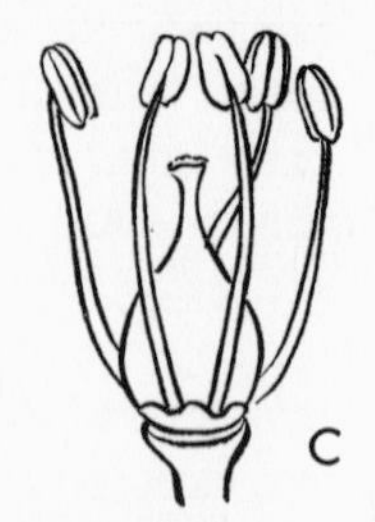

FIG. 8. (A) *unopened flower bud;* (B) *flower bud partly opened, corolla working itself off;* (C) *the act of self-pollenization.*

Once the corolla has worked itself off and the stamens stand erect, it is not long before the pollen sacs on their ends split open, the grains of pollen are released and some of them, with the help of gravity, insects, and wind, are deposited on the stigma, and the fertilization of the flower is finally achieved when a cell from the pollen grain is fused with the egg cell lying far down in the ovary.

This is "normal" fertilization under the conditions of cultivation. But so far as wild grapes are concerned it is highly

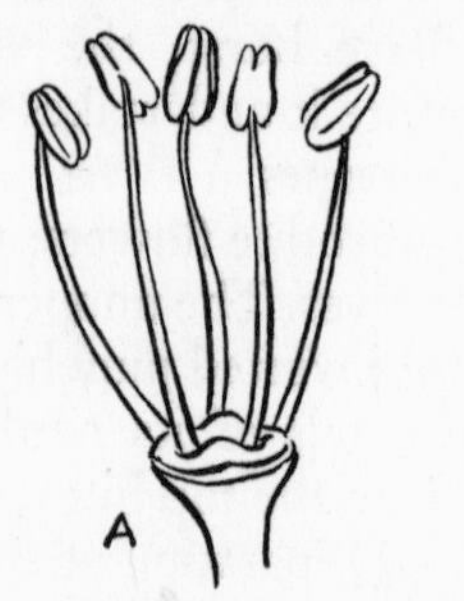

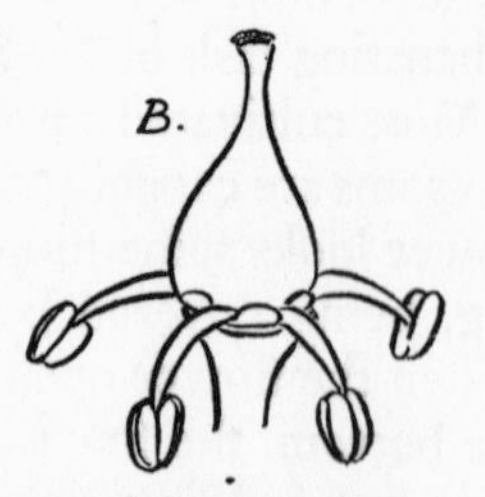

FIG. 9. (A) *male flower;* (B) *flower with reflex, or defective, stamens.*

abnormal. For among wild grapes, the sexes are usually separated. There are vines that have only male or staminate flowers, and there are vines that have only female or pistillate flowers. The staminate vines cannot produce fruit, because their blossoms lack the female organs in developed form, but their pollen may fertilize the flowers of nearby female vines. The wild female vines cannot bear fruit unaided because their stamens are rudimentary or defective. From Nature's point of view, this separation of the sexes is all very well, for it insures cross-pollination and hence the maintenance of the vigor of the species. But for man, to whom the fruit is something to eat or make wine of and not just a younger generation of grapevines, it is inconvenient. For if he seeks to domesticate such a vine he must cultivate indolent male vines along with the females in order to get any crop at all.

One popular misconception may be cleared up at this point. *Cross-pollination does not change the character of the resulting fruit.* The flowers of a Concord vine may be fertilized with pollen from any other variety: its fruit will still to all intents and purposes be Concord grapes. *Only the seed is affected, and the effect of cross-pollination is revealed only if the seed is planted.*

Most cultivated varieties, then, are self-fertile. Unfortunately, a number of the hybrid varieties of the wild American species are self-sterile and hence must be fertilized by staminate vines. In addition to the completely self-sterile varieties, there are also quite a few, including a number of the *viniferas*, which are only partly self-fertile. Before setting out a vineyard, it is necessary to know the fruiting habit of the variety one proposes to plant, so as to interplant with staminate vines or with strongly self-fertile vines of another variety if it is necessary. All the varieties described in Chapters XII and XIII, as worth growing are self-fertile.

Fruit. The fruit of the grapevine, the ways in which it varies as between one variety and another, and the role of its principal parts – skin, pulp, and seeds – was discussed in some detail on pages 7–12. Here, two or three other points may be added. A grape berry is the ripened ovary of a single grape flower. A bunch of grapes consists of the ripened ovaries of all those flowers composing an inflorescence which are successfully pollinated. Self-fertile varieties, as we have seen, usually have a good "set" of berries. The self-sterile varieties, which must depend on stray pollen from other vines, frequently "set" poorly in ragged bunches: their set can be improved by hand pollination — that is, by brushing or shaking pollen from other vines over them. But this is not a commercially feasible operation.

Leaves. The leaf consists of the leaf proper, called the *lamina* by botanists, and the stem, or *petiole.* At the point where the stem meets the leaf itself, it is divided fanwise into five principal veins or nerves, and these are divided and subdivided into a complex system of nerves, never identical in any two leaves. These nerves are in effect a continuation of the transportation system of the vine's root and stem; and they bring the crude sap into the leaves to be elaborated, and carry it back to the growing parts of the vine.

The shape of the leaf differs a good deal according to variety. Deep sinuses may separate it into five lobes, or three lobes; or the leaves may be entire — that is, without strongly apparent lobes at all. The leaves of some species and varieties are woolly on the under side; others are smooth. Some are coarse and leathery on top, and others are very smooth, thin, and shiny. The dentation around the edge of the leaf shows fairly consistent patterns for different species and varieties. And the *petiolar sinus* (the major indentation where the stem joins the leaf proper) may be V-shaped or U-shaped, and either shallow or deeply cut. The shape of the petiolar sinus is used to distinguish many species and varieties; and so is the angle that the two basal nerves form with the petiole.

So much for the leaf's form. Its principal function, which has already been mentioned, is that of elaborating the crude sap brought up from the root-hairs. In addition, the leaf constantly exudes the excess of water that is needed to carry the nutrient material in solution; if this process is interrupted, the vine is unable to breathe, the circulation of water within the vine is stopped, and there is no means of carrying the elaborated food to the various parts of the vine that require it. For these reasons a vine must be well supplied with healthy foliage. Injury to leaf is injury to fruit. When

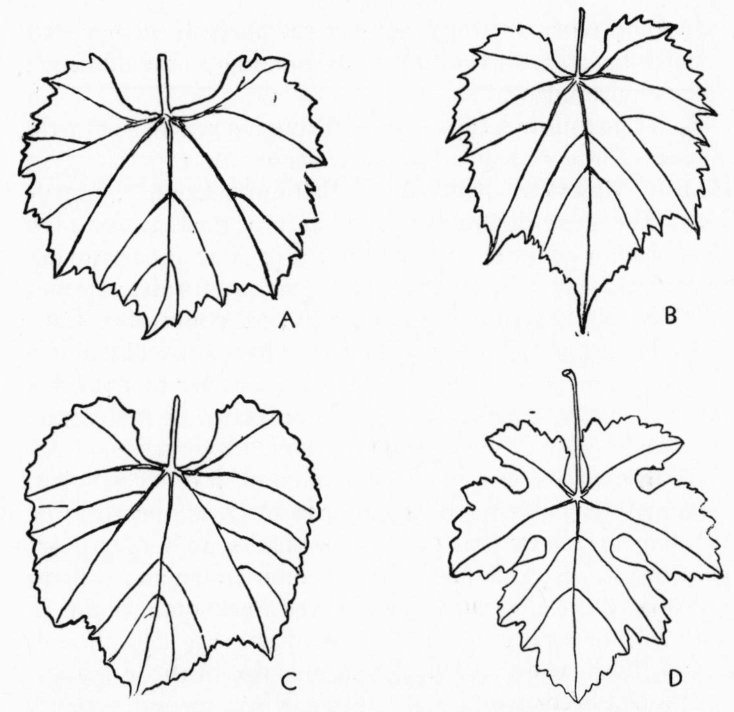

FIG. 10. *Typical leaf shapes of four important species. Note characteristic difference in form of petiolar sinus, dentation, lobing or lack of lobing, length and breadth, and distribution of five principal veins. Characteristic texture and network of lesser veins are not shown in these drawings.* (A) *rupestris;* (B) *riparia;* (C) *berlandieri;* (D) *vinifera.*

foliage is removed, or when it is attacked and injured by insects or disease, the vigor of the vine is affected and its life may even be threatened.

2. HOW THE VINE GROWS

The nature and functions of the parts of the vine are thus summarized. Yet the most minute and careful description of that kind will convey no very accurate notion of the annual cycle through which the vine passes, of the vine as a living and sensitive thing, subject to spurts of energy and fits of exhaustion and constantly responding to the influence of its environment.

So let us follow a typical vine through a season's growth, a vine of the red-wine variety, Baco No. 1 growing in Boordy Vineyard, Riderwood, Baltimore County, Maryland. The same description will do, in a general way, for the behavior of other hybrids French or American growing in central New York State or in Texas, and for the varieties of *V. vinifera* growing in the cool Coast counties of California, in the blazing heat of that same State's southern desert lands, in the inland valleys of Utah, or, for that matter, in the Russian Crimea, or the Andean plains of Argentina. Every variety and every species has its idiosyncrasies, and these in turn are profoundly modified by local conditions. One may have the habit of budding early, and another of budding very late. One may be satisfied with a very short growing season, and another may require many months to grow in. Others, as already observed, may quickly die in limy soil, or swampy soil, or loam that is too rich. But in spite of exceptions and qualifications, the story of the annual cycle of a specific vine growing in a specific place is essentially the story of them all.

Let us say that it is a pleasant day in mid-March, and the signs are favorable for tillage. The vine-grower goes out to do some pruning before he starts on that, in order to avoid having to wade around later in the soft, freshly tilled ground. The buds have not yet started, and to all appearances the vines are still dormant. He prunes the Baco vine, and promptly, at the end of each freshly pruned cane, a colorless tear is formed, is dropped, and is followed by another. Some call this weeping, and some call it bleeding. The tears of a vine freshly pruned in the spring are exceedingly copious, and invariably move the tender-hearted vine-dresser to sympathy. The classic experiment of Stephen Hales, made in 1725, demonstrated that the tears are not only copious but that they have considerable pressure behind them. He attached a tube to the stump of a freshly cut vine, and in a comparatively short time the sap in the tube had risen some 23 feet. More precise measurements of the sap pressure have since been made.

This early sap consists mainly of water, and contains only about 2 parts in 1,000 of plant nutrients. The nutrient material consists partly of mineral substances extracted (with water) from the soil, and partly of organic compounds that have been stored in the vine during the dormant season. The sap starts to flow three weeks or a month before the buds of the vine actually begin to "push," for the young root-hairs begin to develop on the roots during late winter and are already at their work of absorption. A freshly cut vine may discharge as much as a quart of sap in a day, and during a single spring before the sap finally stops pouring from the wound it may discharge a gallon or more. Some investigators contend that the traditional view is correct – that the loss of all this early sap is bound to injure the vine in some degree; yet the actual quantities of nutrient material that it contains are so small that not much is lost, and the prevailing opinion today is that the loss does no harm whatever. And even though it may cause some slight weakening, there are circumstances under which the advantages of late pruning far outweigh any such hypothetical injury. The Baco vine, for example, has the habit of pushing its buds early, and the young shoots run a certain risk of being killed by a latish frost. But since a spring pruning delays budding, this Baco vine in Maryland is never pruned until after the sap has started to run.

On April 1, the warm weather having continued, an examination of the Baco shows that the buds have swelled somewhat: they look alive. On April 23, the two glossy brown scales that cover each fruit eye have begun to part. Apples, cherries, and dogwood are all in blossom.[1] On April

[1] Farmers set planting and cultural dates less often by the dates of the calendar than by certain natural phenomena, of which the blossoming of apple trees is a very important one. Poison-ivy blossoming is another. The dates here used apply to Boordy Vineyard only. Farther north the apples and cherries blossom later, and the growth of the vine is correspondingly delayed. Farther south, of course, the cycle begins earlier.

25, the scales have parted, disclosing the brown cottony covering of the buds. On the following day the shoot begins to emerge, bronze-green in the case of this variety, the growing tip and the first leaf still tightly folded. On April 27, the more advanced buds are already showing their first small inflorescences. This period from the opening of the bud to the appearance of the first inflorescence on the young shoot is called by the French (who naturally, in view of the part the vine plays in their lives, have a minute and comprehensive terminology) the *débourrement*. A prediction of frost for the night of April 29 causes some mental anguish; but the vines, being trained on their trellis, are not affected by this ground frost.

The manner of the young shoot's unfolding has already been described, and these young Baco shoots behave in the immemorial fashion, though during the first few days they are held back by much cold, wet weather. But by May 14 many of the shoots are 14 inches long, all of the inflorescences are showing, and each inflorescence is well differentiated into its dozens of individual flower buds. On May 27, many of the Baco shoots are four feet long, and growing with great vigor. The last of the frost season is two weeks behind.

On May 29, between five and six weeks from the beginning of the *débourrement*, the more precocious inflorescences on the Baco shoots begin to blossom. This is about the average interval for Baco; sometimes it is longer, sometimes a bit shorter. By June 6, four days later, about half of the Baco blossoms are blooming. By June 8, about two-thirds of the blossoms are open, and a good many of the stamens have begun to wither away, indicating that the transfer of pollen from stamen to stigma has been achieved. By June 12, the blossoming is practically complete, and on June 16 – about two weeks after the blossoms began to open in numbers – the entire period of fertilization is over. In some years, if there is plenty of warmth and sunshine, the blossoming period will be shorter; in cold, damp weather it will be considerably longer and much less satisfactory. The ovaries of those inflorescences which blossomed earliest are already small, hard berries. The blossoms that were not successfully fertilized have withered and dropped. The inflorescences have now been transformed into long, slender bunches of tiny young grapes. The grapes at this stage bear a strong chemical resemblance to the shoots on which they are carried: they are green with chlorophyl, and like the leaves they are busy at the work of carbon assimilation. A chemical analysis indicates that they actually contain less sugar than the leaves of the same shoots on which they are growing.

The period between June 16 and July 27, for this Baco vine, is biologically uneventful.[1] At first, the shoots continue to grow with great vigor, but presently this growth slows down and the shoots begin to harden. Simultaneously the leaves start to lose their youthful aspect, to thicken and grow darker in color. The grape berries themselves grow steadily in size. This growth is not a multiplication of cells, however, but a mere expansion of the already existing structure. The chemical composition of the berries does not change much with their growth during this period, although they gain somewhat in acidity.

July 27 is a vital date in the history of this vine's year. For on July 27 the Baco grape-berries show their first faint flush of color. This indicates the beginning of the period which is known by the French as the *véraison*, and for which there is no comparable English word. And the aspect of the vine on this date goes to show that it is entering upon a climacteric period: it droops, it has a weak and, for the perceptive, a weary appearance. This is the beginning of the ripening. The berries have now attained their full quota of acidity and have reached almost full size. Coincident with

[1] But culturally it is exceedingly eventful, for grapes in general, for it is precisely this hot and often humid period of early summer that offers to young grapes and the growing canes the most serious risk of attack by insects and by the various fungus diseases to which they are subject; and it is this period of rapid weed-growth that also requires most of the cultivation. In Maryland, also, it is the period of greatest activity for the bothersome Japanese beetle.

the first appearance of color, they begin to soften (and white varieties appear translucent). The vegetative growth of the vine is practically finished: the *aôutement* (literally, the "Augusting") of the shoots gets under way and they begin to look woody, although the coming of much rainfall in late August and September sometimes starts a new shoot growth at the tips. The leaves of some varieties begin to show color. From this time until the fruit is ripe, the acidity of the grapes decreases steadily, there is an increase of sugar (dextrose and levulose, or grape sugar, mainly, though in some American hybrids there is an appreciable quantity of sucrose also), and the berries absorb water. By August 15, the Baco grapes appear to have attained full color, and the grapes are appreciably sweet to the taste, but still too tart. On August 20, they are very juicy. This is a trying time for the fruit, for as they soften they grow subject to mold, particularly if the weather is hot and wet; and the birds are after them, and the wasps; and when a wasp punctures a berry the bees invariably follow to gorge on its contents. At this time, also, small boys begin to show a marked interest in viticulture.

A close examination of the Baco vine shows that the shoots have been putting forth laterals from the axils of the leaves, that a few of these laterals have borne inflorescences, and that these inflorescences have flowered unseen and are now little bunches of very green, hard grapes. No matter – these second-crop bunches will never ripen. A close examination of the shoots also indicates, at the axils of the leaves, the developing eyes which will yield next year's crop of shoots and fruit.

On August 22, a few grapes are picked and crushed, and their juice is tried for sugar content. It shows 18% of sugar by volume and 15 gm. per liter of acid. The acid is still too high for wine, and the sugar is too low. The grapes, in a word, are not yet ripe.

On August 28, some more grapes are picked and tested. This time they show 20.6% sugar and 10 gm. per liter of acid. The stalks are beginning to be fragile and to snap easily.[1] The analysis shows that the grapes are sufficiently *ripe for wine* (though they would be better if sugar were a degree or two higher and acid a gm. or two lower), and they are picked; any more waiting would mean a further reduction of crop by birds, insects, small boys, and the other animate enemies of the vine, and the possibility of injury by rain.

But though the grapes are ripe for wine, they are not yet biologically ripe. That condition is reached only when there is no further absolute increase in sugar content. After that point has been reached, there may be a further increase in *apparent* sugar content, which is caused by the evaporation of water and a consequent concentration of the juice.[2] Perfect ripeness, from the point of view of the vine, is probably not attained until all the water of the juice has evaporated and the flesh of the grape has raisined, surrounding the dry seeds with withered pulp.

The growth of the Baco vine continues for some time after the grapes have been gathered, because the Baco is a fairly early-ripening vine (it was a week later than usual this year) and there is still some time before the dormant period sets in.[3] But growth activity decreases steadily. Already the thin layer of cork beneath the surface of the shoot has begun to form, cutting off the supply of food from the outer layer and from the leaves. The cane begins to turn yellowish, from the base of the shoot outward toward its tip, and grows steadily darker in color; if late growth has been induced at the ends of the shoots by unseasonably warm weather and plenty of rain, this new growth remains green

1 The stalks of other varieties lignify at ripening and will not snap off, but must be cut with knife or scissors.

2 In districts where the climate is favorable, it is quite usual to allow the grapes of certain varieties to shrivel on the vine this way in order to secure a concentrated must, or juice, from which very sweet liqueur wines are made. And in some districts, especially the Sauterne district of France, the drying-out process is helped by the development of a mold on the grapes, *Botrytis cinerea*, the "pourriture noble," which absorbs much of the water of the grape even as it hangs.

3 This variety is really too early to be ideal for Boordy Vineyard. But it does well enough.

and never does ripen properly. As the shoot matures into woody cane, the vine shifts its activity from the promotion of further growth to the storage of starch and other food, reserves which will be used by the vine in the initial burst of growth the following spring.

On September 30, most of the Baco's canes have ripened satisfactorily; the leaves are still green, but they begin to look withered, a sign that their work of carbon assimilation is almost over, that the vine is now amply supplied with reserves of food.

On the night of October 10, there is a heavy frost, 18 days earlier than the average date of a killing frost in these parts. The work of the vine for the year is abruptly ended, and gradually during subsequent days the withered leaves drop to the ground. It was an abnormal year, with a cold, moist spring, a moist summer, less hot than usual, a cool, moist fall – altogether too much moisture – and a growing season cut short by the frost. But then, vines are used to abnormal seasons, for every season is abnormal; and after all, the vintage was not reduced by any major disasters and the grapes achieved a satisfactory if not perfect maturity. The vine did its job well, all things considered, and perhaps it deserves its extra two weeks of sleep. Below is a table showing to what use the vine put the season:

Débourrement to blossoming	35 days
Blossoming period	15 days
Blossoming to *véraison*	46 days
Ripening period	32 days
Total	128 days
Normal growing season, this vineyard	185 days

The dormant period, into which the vine now lapses with the shedding of its leaves, is necessary for the production of good grapes. When grapevines are grown in tropical or subtropical regions, which have no winter, they become evergreens, and this unaccustomed evergreen habit lures them into fatal inconsistencies; shoots are being put forth, flowers are blossoming and fruit is ripening all at the same time, with the result that there is a small continuous production of worthless fruit accompanied by a tremendous burst of vegetative activity, and (the usual penalty of excess) an early death. That a period of rest is for some reason necessary is further shown by the experience of those who raise grapes in hot-houses. No matter how strongly the heat is turned on, hot-house vines will resolutely refuse to grow at all, once they have lapsed into their dormant period, until they have had a certain minimum period of rest.

When a vine lapses into dormancy, what it does biologically is to seal itself against the loss of water. It will be remembered that one of the chief functions of the leaves is to give off the vine's excess water supply; hence the first step toward dormancy is the shedding of leaves. The lignification of the canes, the development of the cork layer beneath the bark, likewise help to seal against the loss of water.[1]

Hence the hardy vine is the leak-proof vine, the vine with the ability to hold its moisture. The practical conclusion to be drawn from this is that unless the vine's wood is properly ripened it is likely to be killed by loss of water.

So the Baco vine, having yielded its fruit, lapses into the dormant state. And there it remains, first in a deep rest from which no amount of Indian summer can stir it, and then, toward the end of winter, into a troubled sleep from which too much unseasonable warmth may sometimes rouse it, but which normally has no other effect than the stimulation of the roots to put forth new rootlets in preparation for the demands of another season.

[1] But, as we have seen, the capacity to seal itself off sufficiently to survive intense cold varies greatly from one variety to another.

Chapter VII

ESTABLISHING THE VINEYARD

1. FIRST STEPS

ON THE banks of the Rhine, the establishment of a new vineyard has often in the past meant virtually the rebuilding of a hillside. Topsoil was carefully removed, and the subsoil turned over to a depth of several feet. Rock, if it was of the type that decomposes into good vineyard soil, was blasted, broken, and exposed to the weather. The hillside was laid off in an elaborate system of terraces, broad or narrow according to the steepness of the slope. A system of paths and roadways was worked out. Terraces were buttressed with stone walls. Frequently the terraces were dressed with new topsoil.

All this was justified, because the vineyard owner knew that the right grapes grown in the right locations along the river would yield wine capable of bringing extravagant prices. But nowhere in the United States is any such lavish preparation justified. In California, it would be commercial suicide. For though the vine is cultivated on a larger scale in California than anywhere else in our country there are still thousands of acres of good vineyard land that is not initially expensive and does not call for any such fancy preparation. In other parts of the United States – with one significant exception – a large initial investment is even less justified, for commercial wine-growing in the East is still by and large an untested branch of agriculture. It may in time be quite profitable, should a taste for table wines be properly nursed along; but it is still largely experimental as regards both the market and the adaptability of the various regions. The exception is the wine-growing region of the Finger Lakes. There, where most of the good grapes go into sparkling wine, wineries can afford to pay high prices for their grapes simply because the cost of the material is a relatively minor item in the making of sparkling wines. An expensive location, on which maintenance costs are high, is therefore justifiable in this region.

As steps in establishing a vineyard, the prospective wine-grower should first discover whether there are any vineyards in his locality, what grape varieties are grown, and what cultural problems face the local grower. If he plans commercial production, he must also investigate the State and Federal laws relating to wine-making; this is, of course, not necessary if he plans to make wine for his own use only. Then he ought to find out something about the local climate — length of the growing season, summer temperatures, winter temperatures, frost dates, the amount and distribution of rainfall, humidity and prevailing winds — and compare these facts with the part of Chapter V that has to do with his region. He must decide, with the help of the information in Chapters XII and XIII, what grapes to grow. Then he must find a nurseryman who deals in these varieties. Once he has located the source of his vines, he may begin to concern himself with the preparation of the vineyard site.

How large shall the vineyard be? In some European districts, a vineyard of four to six acres yields a modest living for its owner. In California, on the other hand, there are profitable vineyards of thousands of acres. In the established vineyard regions of the East — in the Chautauqua region, to be specific — 30 acres was considered about right for horse culture, and 70 vineyard acres is a good size for a single tractor. But the prospective commercial wine-grower in regions that are not well established will make a modest beginning, perhaps starting with a limited planting of several varieties and adding to it only as he finds his market and determines the grape varieties best adapted to his requirements. The amateur will be more modest still. If he proposes to do the work himself in his odd moments (with some occasional help from a hired man) he can, if he likes exercise, take care of a half-acre without difficulty and produce from it enough wine to keep his family and all his friends perpetually afloat. For all save the most ambitious and enthusiastic amateurs, a vineyard of a half-acre is actually too large, since after the first vintage or so the reserves of wine begin to pile up at a great rate.

It is easy for the amateur to estimate in advance the approximate yield to be obtained from a given piece of vineyard, and to lay his plans accordingly. Four facts must be taken into consideration:

1. A ton of grapes yields about 150 gallons of wine.
2. The hybrids and fine-wine *vinifera* sorts will run 4 to 15 pounds of grapes per vine per year, and the bulk-producing *vinifera* somewhat more.
3. The native hybrids are usually planted 8 feet apart in the row, and in rows 8 feet apart: 64 square feet per vine. This area is about right for *vinifera* also.
4. There are 43,560 square feet in an acre. An acre is roughly 220 feet square.

From these facts, computation is easy. Assume that an amateur has a suitable site 80 by 80 feet. On this piece of ground he can set out 10 rows of 10 vines each, or 100 vines. Annual yields will vary a good deal, but assuming 8 pounds per vine the annual yield will amount to 800 pounds, or two-fifths of a ton – enough to produce around 60 gallons of wine a year. By cheating himself a little,[1] he may stretch this yield considerably.

2. PREPARING THE SITE

The preparation of a site varies a good deal with its location, its soil, and the use to which it has previously been put. A very steep slope is best avoided; if it cannot be, then it had

[1] By "ameliorating" the must. See Philip M. Wagner: *American Wines and Wine-Making*, p. 188.

better be terraced if serious erosion is to be prevented. A gentle slope is easier to prepare and easier to manage. A piece of level land, if it has good drainage and is not too much subject to frost, will do. A wet site must be drained, and drainage is an expensive operation; commercially, it is seldom worth while to use for a vineyard a site that requires draining. If a site has not been cultivated before, trees and brush must be cleared off, roots grubbed out, and large stones removed.

In Europe, the site for a new vineyard is first given a thorough trenching. This fiendishly difficult job is usually let out to contractors. It consists essentially in plowing both the topsoil and the subsoil without mixing the two, or (under special circumstances) of making the topsoil the subsoil and the subsoil the topsoil. The idea is to provide a deep and permeable bed for the roots of the new vines. In Europe, trenching is frequently carried down as far as three feet, but there are compromises. One is to plow a furrow, and then, behind the plow, to loosen the subsoil with picks, then to turn the next furrow over the loosened trench, make another attack on the next exposed strip of subsoil, and so on. But this method requires considerable hand labor and is too expensive. The next best compromise is to plow and to follow the plow with a subsoil plow which stirs the ground several inches deeper without turning it. Still another compromise is to plow the site into 8-foot "lands" – that is, the width of the prospective vineyard rows – and to run a subsoil plow through each dead furrow at the 8-foot interval where the vine rows will be planted. But most grape-growers are satisfied to choose in the first place a site with a fairly permeable soil, and to content themselves with a good deep initial plowing.

Anyway, a vineyard site should be well plowed a year before the young vines are to be set out, and sowed with some crop. Before sowing, it should of course be graded for irrigation in regions where irrigation is available or necessary. And if the land is run down, it should receive an application of fertilizer, preferably manure, plus a cover crop that may be plowed under in late fall. The site should have a fall plowing, whether it is a cover crop or a "catch crop" grown during the summer.

When the ground is in fit condition in the spring, it should have its final plowing before the vines are planted, or at least a thorough harrowing; for it must be remembered that when grapevines go in they go in to stay for a long time. Spring plowing should not be undertaken while the ground is still wet, for then the soil (especially if it contains much clay) dries in heavy clods that make subsequent cultivation difficult.

3. THE VINEYARD PLAN

A vineyard plan ought to be worked out well in advance, so that the right number of vines may be ordered, and to allow for any future expansion. It should be closely adhered to. The first thing to determine is the space that should be allowed the vines both in the row and between rows. The native hybrids, as they are grown in the northeastern States, are generally planted 8 feet by 8 feet. An 8-foot alley is wide enough for cultivation by hand, tiller or real tractor, though 9-foot alleys give a tractor more leeway, and an 8-foot interval in the row gives space for good development without crowding, and allows room for the circulation of air. Varieties of low or moderate vigor may be planted closer in the row – say, 6 feet. Farther south, where the vines tend to be more vigorous, a 10-foot interval in the row is better, and for ultra-strong-growing varieties 12 feet is still better: there is not much diminution of crop, because the larger vine is more productive.

The growing habits of the *vinifera* differ from those of the native hybrids. They are stocky by habit, whereas the native hybrids run to length and leanness.[1] Consequently, *vinifera* may be planted under some circumstances as closely as 4 feet apart in the row. In the Burgundy and Champagne districts, the vineyards used to be planted *en foule* (helter-skelter), and so closely that 8,000 or 10,000 vines, each

[1] The French hybrids are about equally divided between those that grow compactly like most *vinifera* and those of rampant growth.

pruned back to one or two eyes a year, struggled for survival on a single acre; now wire trellising is almost universal in these districts. Planted 4 feet by 4 feet, an acre will hold 2,722 vines. But the hot climate of California, and the requirements of mechanical cultivation, will not allow such crowding. Winkler recommends for California a minimum of 72 square feet per vine (6 feet apart in the rows, 12-foot alleys) for cane-pruned and trellised varieties, and 64 square feet (8 feet by 8 feet) for the head-pruned varieties that do not require a trellis.

Assuming a fairly level piece of land, the best plan is rectangular. This allows cross-cultivation for all varieties up to the time when the trellis is erected, and for the life of head-pruned, or untrellised, vineyards. It is important to remember that vines should not be planted out to the very edge of the site. For a trellised vineyard, a strip of 8 or 10 feet must be left on each side for cultivation, and turning space at the ends of the rows. An untrellised vineyard, allowing cross-cultivation, requires turning space at the sides also.

It is important in large vineyards to break the rows every 200 to 250 feet by transverse avenues 16 or 18 feet wide (wide enough to let vehicles through). Otherwise the transportation problem will be needlessly complicated. Quick and easy access to any part of a large vineyard may be obtained only when it is cut by avenues at regular intervals and is surrounded by a roadway.

A trellised vineyard in a cool region is perhaps a bit better off if its rows are placed north and south, for the sun at midday can then shine directly down the alleys. In hot climates, an east-west direction is better, because it helps (a little) to shade the vines during the hottest part of the day and thus to minimize danger of sunburn. But sometimes the direction of the rows is not a matter of choice. In a short-season area, a southern exposure provides appreciably more heat, and best of all is a sheltered pocket facing south. When irrigation is necessary, the direction of the slope dictates the direction of the rows. The question of fitting the vineyard into the prevailing irrigation arrangements is a matter of such local and specialized interest that it will not be entered into here. And when the vineyard is placed on a slope, there is likewise no choice. Rows should follow as nearly as possible the prevailing contour lines in order to minimize erosion. Contours cannot, of course, be followed with strict accuracy; for on most slopes they would twist and turn in a manner to confound the most careful cultivator, and there is the further difficulty that a trellis that twists and turns is very expensive. Two end-posts only are needed for a straight trellis, but a curved trellis would require heavy posts at frequent intervals. So, in contour-planting a trellised vineyard, the ideal must be sacrificed to compromise.

A vineyard plan must be transferred to the plowed field. Here, if a rectangular plan has been decided upon, the first thing to do is to establish a right angle. The square of the hypotenuse of a right-angle triangle equals the sum of the squares of the other two sides. Establish the corner point, and measure off 24 feet on a length of string. Then with the string lay off a triangle whose sides are 6 feet and 8 feet respectively and whose hypotenuse is 10 feet. That's a right angle. It is then easy to lay off the parallel vineyard rows and with small markers to indicate the points on the rows where the young vines go.

4. PLANTING

Some nurseries offer both 1-year-old vines and 2-year-old vines. If the 2-year vines are more expensive, buy the 1-year. There is little difference between the two in the rapidity of growth after planting—provided they are sound, well-rooted stock. *Never* buy older (the so-called "bearing age") stock. In California, many nurseries deal in ready-grafted vines of the leading wine varieties. Some carry even larger selections of wine varieties on their own roots. Most of the commercial grafts are on Rupestris St. George, which is the most generally satisfactory for California conditions; but by ordering in advance it is sometimes possible to get ready-grafted vines on other stocks. From some, mother-vines of various rootstock varieties may be bought, from which one may in time secure a plentiful supply of one's own grafting wood.

The horticultural expert of your State experiment station can usually recommend nurserymen, but it is best to deal with nurserymen who specialize in grapevines and don't just carry a few vines along with a general selection of horticultural stock. (A number of such nurserymen advertise in the *American Fruit Grower* and in the California paper,

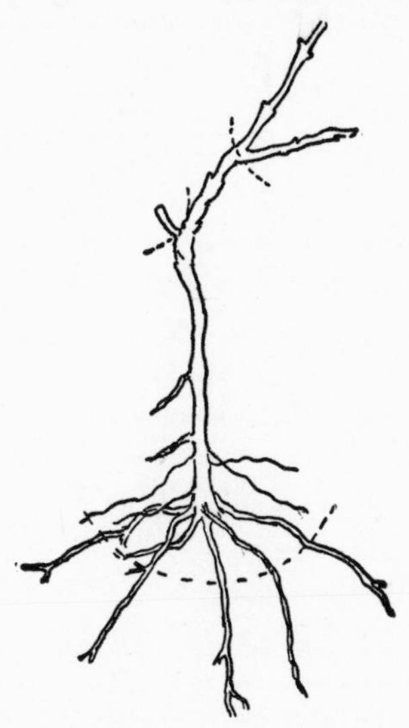

FIG. 11. *Young vine as received from nursery. Dotted lines indicate cuts to make preparatory to planting.*

Wines and Vines.) Prices of vines vary according to demand: in normal times they range from around $25 per hundred for the old standards such as Catawba to $65 or more per 100 for new introductions or hard-to-propagate varieties. In any case, initial cost is relatively low.

Vines ought to be ordered six months before planting; otherwise there may be disappointment. The order should be placed for delivery shortly before planting time. Planting may be done any time during the dormant season. But as a general rule spring planting is preferable to fall planting, since a number of hazards are thereby avoided. The vines usually come wrapped in damp moss and oiled paper to keep them from drying out. They should be "heeled in" immediately – that is, lightly buried – in order to keep them in good condition until planting time. During planting, they should be taken out only as needed, and preferably kept in a pail that has a little water in the bottom.

The work of planting is quick and easy. A hole is dug,

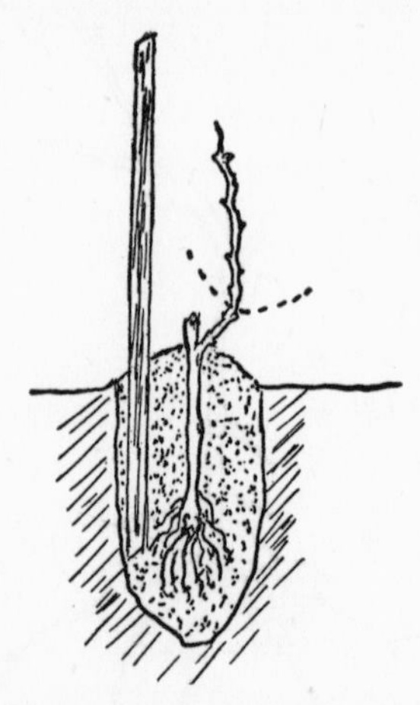

FIG. 12. *Young vine planted, with stake for tying during first year's growth.*

deep enough to accommodate the young vine and about a foot in diameter. A pole or stick about 3 feet long[1] is placed upright in a corner of the hole. Then a young vine is trimmed for planting by cutting the roots back to 3 or 4 inches and cutting the top back to two eyes; it is then placed in the planting hole so that the two eyes protrude an inch or so above the ground, and the roots are spread out in the hole and firmly covered with some topsoil. The entire hole is then filled and the dirt tamped firmly. Thus the vine is planted, and beside it is a support to which the young

1 Ordinary building lath is cheap and adequate.

shoots may be tied when growth begins. Grafted vines should have soil loosely mounded over the protruding eyes in order to protect the graft until the vine is well established.

A family-size vineyard is easily set out in an afternoon.

In setting out large vineyards, the labor should be divided. One man digs holes, and another trims vines, places stick and vine, and fills the hole. Some California vineyards are planted, not by digging holes, but by plunging a dibble into the ground at the right spot, inserting a vine which has had its roots trimmed close, and tamping the ground around it. This speeds the work, and gives satisfactory results, especially if it is possible to irrigate shortly afterwards and thus to settle the soil compactly about the vines. It cannot be done on a coarse gravelly or stony soil. In California, in setting out a vineyard which is to be head-pruned and which will never need a permanent trellis, it is customary to use fairly substantial redwood stakes, 1 to 1½ inches square and 36 to 40 inches long, which serve as supports for several years until the headed vine has a sufficiently stocky trunk to support itself. Stakes should be on the leeward side of the vine to prevent undue strain of the young shoots at the points where they are tied.

Once planted, the young vineyard is given a thorough irrigation where irrigation methods prevail. From then until they have established themselves and the first cultivation is needed, there is nothing to do but watch and wait, hoping that cutworms and frost will be merciful to tender young shoots.

5. THE TRELLIS

Head-pruned vines as grown in California never require a trellis. For vines that are cane-pruned and hence do require permanent support, the trellis need not be put up until the end of the first, or even of the second, growing season.

Trellises may differ considerably in detail – that is, in the number of wires, the size of the wire, and their height above ground. Reasons for these differences will be explained in the next chapter – "Pruning and Training." But essentially trellises are all the same, consisting of end-posts, intermediate posts, and the wire that is strung between them. For the sake of simplicity, only one type will be described—the two-wire trellis most commonly used in the eastern vineyards.[1]

The first step in the construction of a trellis is the setting of the end-posts. These should be 6½ to 7 feet long, 4 to 6 inches in diameter, and of some durable wood such as cedar, black locust, osage, or white oak. Metal posts are even better, but expensive. Second-hand 1½-inch galvanized water pipe does admirably. Locate the position of the endpost accurately by sighting along the lines of vines; and place it about 5 feet behind the end vine of the row. Dig a hole (a post-hole digger does the best job) about 2 feet deep, and put the post in place. It may be placed vertically, or inclined slightly *away* from the row. It should be tightly tamped. It has been said that a trellis is no stronger than its weakest end-post. The profound truth of this can be appreciated only by someone who has not set his end-posts deeply enough, or who has put in end-posts without giving due attention to the question of bracing. A thrust-brace is commonly used—a diagonal wooden brace one end of which is secured against a heavy stake and the other end of which fits into a notch on the end-post and keeps it from heaving when the wires are tightened. Instead of this, one may use a tension wire (No. 10 wire is heavy enough) secured to the top of the end-post and stretched from there down and back to a heavy stake or anchor put well behind the post. This type of brace is cheap, simple, and reliable; its only disadvantages are that you trip over it, and that it sometimes interferes with turning during cultivation if space is cramped.

Intermediate posts are lighter, since they carry no strain. The posts should be 6 feet long (so that they may be sunk 18 inches) and about 2½ inches in diameter. These may be sharpened and driven. Metal anchor fence-posts are excel-

[1] But in the author's opinion a three-wire trellis offers distinct advantages for most grape varieties. This will also be gone into in the next chapter.

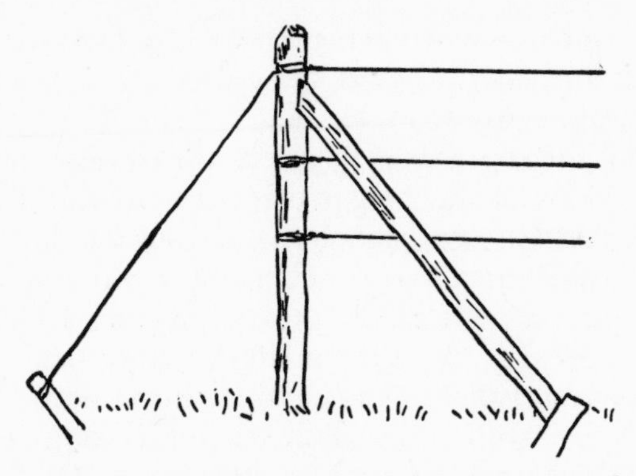

FIG. 13. *Illustrating two ways of bracing an end-post — by thrust-brace or by guy-wire.*

lent (they also serve as lightning rods), and in the long run probably as cheap. Also, they are very quickly and easily driven with an inexpensive device made especially for the purpose. One ought to provide an intermediate post for every two, or every three, vines. Thus, if vines are planted at 8-foot intervals, an intermediate post may be driven every 16 or 24 feet. If the vines are at 10-foot intervals, a post should be driven every 20 feet; if at 6-foot intervals, every 18 feet.

The wires of a two-wire trellis are placed about 2 feet 8 inches and 4 feet 6 inches from the ground. String the top wire first. It should be No. 10 plain galvanized wire, running 2,060 feet per 100 pounds. The lighter No. 12 wire (3,375 feet per 100 pounds) is suitable for the lower trellis wire.

Stringing wire is tricky. An unfastened coil, if left to its own devices, promptly gets itself into a snarl so complicated that it can be unsnarled only with the help of clippers. The simplest way, for a small job, is this: lay the coil of wire on the ground beside an end-post while keeping a foot on the coil, fasten the newly loosened end to the post, and then pay out by rolling the coil down the row to the other end. When the far end-post is reached, the wire is clipped from the roll, pulled tight, and twisted around the top of the other end-post. The wire is then stapled (but not too tightly) to the intermediate posts, at the correct height and on the windward side. Some prefer not to drive the intermediate posts until the top wire is in place, as it is then easy to get good alignment. A slight sag in the wire need cause no worry; it will be taken up by the intermediate posts and by the tension wires of the end-posts. There exist little ratchet contraptions that allow one to regulate tension, which are nice but not necessary. In wiring a larger vineyard, it is best to rig up some sort of reel for unwinding the wire. It is then more simply handled, and all chance of kinking and twisting is avoided.

The lower wire may then be stretched and stapled in the same way.

Chapter VIII

PRUNING AND TRAINING

IT MAY as well be admitted at the outset that this is not an easy chapter to write. Once the essential trick is grasped, the pruning of grapevines is simple and rapid. And on a brisk day in early spring, when the frost begins to relax its grip on the earth and all growing things begin to stir in their dormancy, there are few pleasures more satisfying than a day in the vineyard with the pruning shears, spent relieving the vines of their baggage of excess growth and shaping them to their task of bearing another vintage. No two vines present quite the same problem in pruning, yet all yield to the application of the simple principle and emerge in rapid succession from the vine-dresser's attentions, trim and shapely and ready to begin their growth.

Yet the effort to explain this pleasant business of pruning and training grapevines is surprisingly elusive. Try as one will, the explanation and description seems foggy and difficult to grasp. It is like trying to describe the tying of a bow tie when one has no tie to demonstrate with and one's interrogator is none too clear as to the end in view. Ten minutes with a man who really knows the art of vine-dressing will unveil the whole mystery – it's that simple. But one must be sure that one's instructor really *knows* his business. And failing a competent instructor there is nothing for it but to follow the written explanation carefully with frequent recourse to the drawings and photographs – and with confidence that, as young vines are led through the simple successive steps of training, each successive step will follow naturally from that which has gone before.

1. PRINCIPLES

So to the baffling chore.

A grapevine, left to itself, quickly becomes a tangled skein of woody canes and produces every other year or so a quite large crop of inferior grapes.

A grapevine, properly pruned and trained, grows in a seemly and somewhat less vigorous manner and produces good annual crops of superior fruit.

Since the grape-grower wants orderly growth, an annual crop, and superior fruit (from his point of view, fruit with a sufficient sugar content to yield sound wine), he prunes and trains his vines.

Note the distinction between pruning and training. Training gives a certain preconceived form to the permanent and semipermanent parts of the vine. Pruning regulates the *annual* growth so as to produce the maximum crop consistent with quality and regularity of production. The photographs in Plate V illustrate in a general way the objects of both pruning and training. The photograph at the left shows a grapevine at the end of its season of growth, the crop having been picked long since and the leaves having fallen. The other photograph shows the same vine after it has been pruned and tied and is ready to enter another season of growth – at the end of which it will again resemble the photograph at the left and be ready once more for the pruning shears. In these photographs, the element of *training* is represented by the shape to which each year the vine is restored – which in this case consists of a short permanent trunk about two feet high, from the head of which extend four bearing canes and four short spurs. There are many other methods of training besides this one. The element of *pruning* is represented by the manner in which the mass of woody growth is dealt with so as to get rid of all surplus wood, provide healthy bearing canes of proper length and distribution, and provide spurs for the production, during the year, of well-placed bearing canes for the season after.

So much for a bird's-eye view of the subject. Pruning, its

principles and practice, will now be discussed; then the chief forms of training will be described.

Principles of Pruning. Following are the principles on which pruning is based. They are worth getting firmly in mind, for once they are understood the various systems of pruning follow naturally and reasonably.

1. The crop of grapes has its source in the bud-eyes, which are located on the vine's one-year-old wood – that is, on the canes that the vine produced during the preceding season.

2. Each bud-eye that is left after pruning – i.e., each of the bud-eyes distributed along the four canes of the right-hand photograph in Plate V – puts out a shoot, and the bunches of grapes are borne on these shoots. These grape-bearing shoots grow vigorously throughout the season; and after the crop is gathered from them their character changes – they become woody rather than succulent and become the fruitful canes, or one-year-old wood, of the succeeding year. These are the tangled canes that one sees in the left-hand photograph in Plate V.

3. If very little of this one-year wood is removed at pruning time, and many bud-eyes are left in consequence, there will be a large crop of grapes and a relatively feeble growth of wood. If most of the one-year wood is removed and only a few bud-eyes are allowed to remain, there will be a small crop of grapes but the cane-growth will be vigorous. Hence heavy pruning sacrifices fruit for cane-growth, and light pruning sacrifices cane-growth for fruit.

4. The well-pruned vine is one on which enough – but no more – of the fruitful bud-eyes have been left to yield a good crop of grapes and insure a good growth of well-placed and well-matured canes for the next season.

5. A grape variety that habitually produces its grapes in large, heavy bunches will produce an ample crop from relatively few bud-eyes. If too many bud-eyes are left at pruning time, such a vine will overbear, fail to ripen its crop well, and perhaps be permanently weakened. By contrast, a variety that habitually produces its grapes in small bunches

should be more generously pruned if it is to yield an adequate crop — and may be without fear of overbearing.

6. The *art* of pruning consists in determining, for each vine, how many bud-eyes shall be left and how they shall be placed, the number of eyes to be left depending partly on the inherent vigor and producing capacity of the vine, and partly on its previous treatment. If it has been forced to overbear in the past, for example, the wise course is to prune rather severely, so as to limit the crop and encourage wood growth.

7. Upright shoots grow more vigorously than horizontal or drooping shoots.

8. The eyes of all canes are not equally fruitful. The eyes of exceedingly large and heavy canes and of unusually weak canes are less fruitful than those of canes of "normal" growth.

Systems of Pruning. So much for the principles that govern the fruiting habit of the vine. Let us see how they are applied in the three systems of pruning, called *spur pruning, cane pruning*, and *mixed* or *cane-and-spur pruning*.

Spur pruning, or short pruning, consists in cutting back the fruitful canes to short stubs, or spurs, of two or three eyes each. (See "before and after" photographs in Plate VI.) It is quickly done, and it prevents the vines from rambling and keeps them fairly compact. But it can be used only for those varieties whose basal eyes are fruitful, and, as we shall see when we come to a consideration of training systems, it is applied chiefly, but not exclusively, to heavy-bearing *vinifera* varieties.

Cane pruning, or long pruning, consists in leaving a few fairly long canes to supply the vine's annual quota of fruitful eyes, and removing the rest of the canes entirely. The number of eyes left is not necessarily fewer than it would be if the same vine were to be spur-pruned. The vine is left with two canes, let us say, of 9 eyes each (total, 18 eyes) instead of six spurs of 3 eyes each (total, 18 eyes). Cane pruning has the advantage that it leaves the more fruitful eyes — those located fairly well out from the base of the cane. It has the important disadvantage that, since the end eyes of a cane always make the longest growth, the new wood is after several seasons far removed from the trunk of the vine and is inconveniently placed. Hence cane pruning is seldom used in its "pure" form.

Mixed pruning, as its name suggests, is a combination of spur pruning and cane pruning, intended to embody the best features of each. The essence of mixed pruning is that the unit is not the spur or the cane but *a pair* consisting of a spur *and* a cane. Let us see how one of these pairs functions in practice.

Figure 14 illustrates the principle. In this illustration, Fig. 14A is a young vine from the head of which (we shall say) only two canes, *a* and *b*, were allowed to grow. To form a cane-and-spur combination, they are pruned as indicated by the dotted lines, and the cane *b* is tied to the trellis wire. The pruned vine is then as represented in Fig. 14B. In the course of the subsequent growing season, bearing shoots are put forth by the various bud-eyes on the cane-and-spur combination; and at the end of that season, after the leaves have fallen, the vine's equipment of wood is as represented in Fig. 14C.[1]

And now the rationale of the spur becomes evident. For it will be seen by reference to Fig. 14C that the spur, as well as the fruiting cane, has produced new canes – two of them. *These two canes are destined to be the cane-and-spur combination of the subsequent year.* For at pruning time it is necessary to make only the three cuts indicated by the dotted lines in Fig. 14C in order to put the vine in shape for another crop. The previous year's bearing cane, having served its purpose of yielding a crop, is removed entirely by cut *ab*. A spur for the coming season is provided by cut *cd*, and the new bearing cane is shortened to a proper quota of fruitful bud-eyes by cut *ef*. The vine is then as represented in Fig. 14D – compact and orderly and providing assurance

[1] These drawings are highly diagrammatic, of course. Actual growth is not nearly so regular and orderly as the drawings may suggest.

FIG. 14. *Illustrating mixed, or cane-and-spur, pruning.*

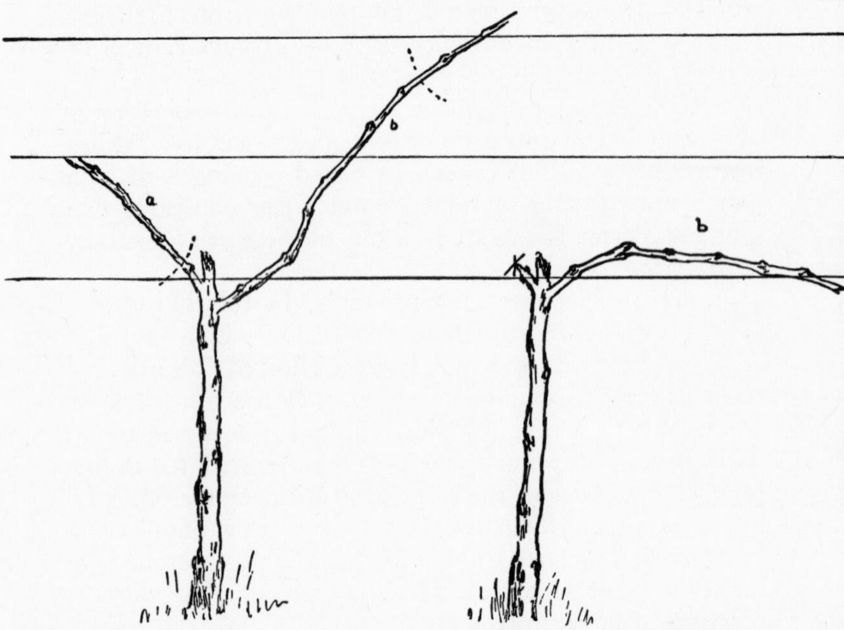

(A) *Young vine with two canes.* (B) *Same vine after pruning to one bearing cane and one spur.*

(thanks to the spur) of well-placed renewal wood for the season after. That is the essential principle of mixed, or cane-and-spur pruning.

2. PRELIMINARY TRAINING

The elements of pruning, as just set forth, apply to all grapes, both *vinifera* varieties and native hybrids. Likewise, all vines receive approximately the same treatment during their first two years in the vineyard, *regardless of variety and regardless of the training system that has been determined upon.* The object of this preliminary treatment is to establish the vine's permanent trunk.

The vines are planted with two eyes above ground, or,

Young shoots (Delaware) at the age of two weeks. Note developing inflorescences.

Close up of young shoot (Seibel 6905) at six weeks. Inflorescences are well developed and are just about to begin blossoming. Mottling on shoot and leaves is spray residue.

Inflorescence (Delaware) in full blossom and already partly pollinated.

The ripening fruit, Seibel 7053.

Pouilly vineyards on steep slopes near Macon, France.

Not all fine wine vineyards are on hilly land.

Vineyard in the French Champagne district, near Epernay.

Dormant vine (S. 10878) before pruning.

Dormant vine (S. 10878) after pruning to four canes and four spurs on a three-vine trellis.

Dormant vine (S.V. 5276) before pruning.

Dormant vine (S.V. 5276) after spur pruning.

Classical looped cane pruning of Rieslings, as used in the Moselle-Saar area, shown just as buds are opening in May. Scharzhofberg vineyard (Egon Muller).

Same Riesling vineyard in the Saar valley, looking up the slope. Note overhead irrigation.

Young dormant vine (Baco No. 1) being trained according to the Kniffin system. Photograph has been retouched to eliminate trellis wires and distracting background. Note two canes and two spurs at the lower wire level and two canes and two spurs at the upper wire level. A year later it will be pruned to two canes and two spurs at the lower wire and four canes and four spurs at the upper wire.

Cane-and-spur training without trellis in the French Beaujolais region – buds just beginning to open.

Method of pruning a French hybrid to canes and spurs on trellis, in the French Beaujolais region.

Pruning and tying in a French white-burgundy vineyard.

Growth beginning in French head trained vineyard, Chateauneuf du Pape.

Old style training in the Rhone Valley of France, Hermitage.

French mass production vineyard on alluvial soil, trained to short cordons, near Arles in the Rhone Valley.

After pruning, but before spring plowing, in the French Burgundy region.

New vineyard in upland Maryland, photographed in mid-May after April planting.

Same vineyard in mid-May four years later, with full crop in prospect.

Vineyard in Napa Valley (Inglenook).

Finger Lakes vineyards, from the ridge overlooking Keuka Lake in central New York.

The ripening fruit, Sauvignon Blanc.

The ripening fruit, Sémillon.

The ripening fruit, Seibel 6339.

The ripening fruit, Seibel 10096.

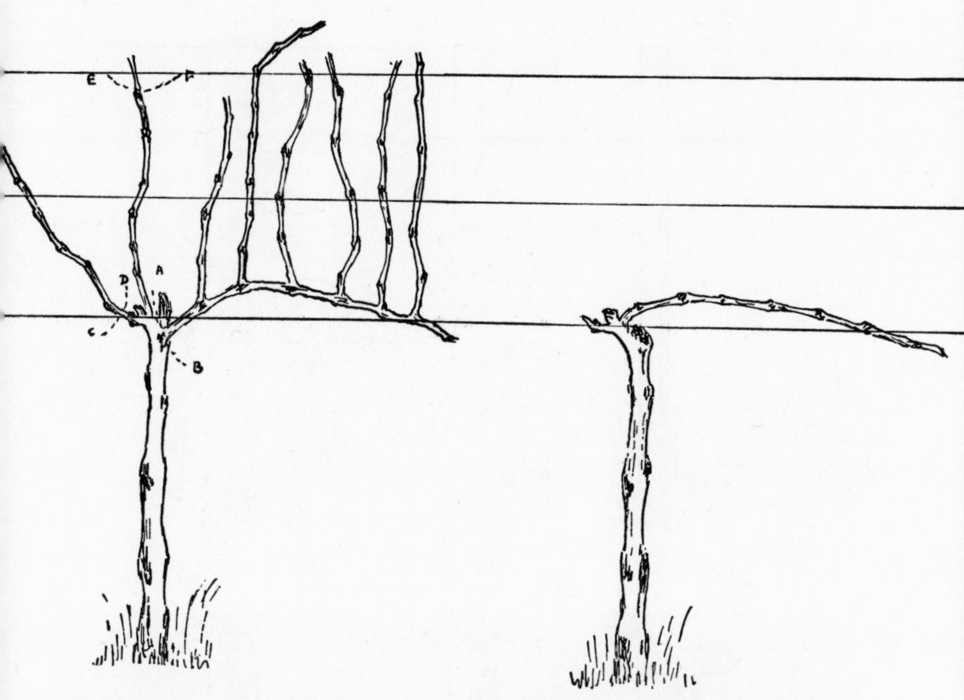

(C) *Same vine at end of subsequent growing season, showing growth of new wood.*

(D) *Same vine pruned again to single cane-and-spur combination and ready for another season of growth.*

if they are grafted, in a loose mound of earth. Shoots emerge from the two eyes and begin to grow. The weaker of these two shoots is eliminated after frost danger is over, and the remaining one is tied to its stake as illustrated in Fig. 15 and tied loosely so as not to injure it. In the course of the season, two or even three ties may be made in order to keep the young shoot straight.

Treatment before the second season's growth begins depends on the amount of growth made the first season. If it has grown vigorously, the cane may be cut off at the proper height to serve as the vine's permanent trunk.

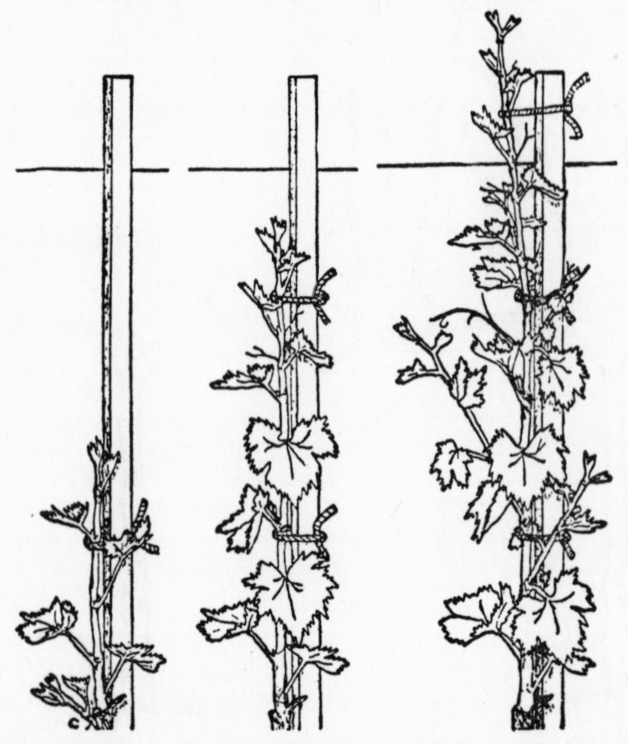

FIG. 15. *How to manage the young vine during first season's growth* (*Cal. Ext. Circ.* 277).

If – and this is more usual – the young vine makes only moderate growth the first season, then before the second season's growth begins it should be cut back to two eyes again and treated as it was the first year. The growth of the shoots will be much more vigorous the second year, and after the weaker one has been rubbed off that which is left will always make enough growth to provide a good sound young trunk of the proper height.

Occasionally, feeble-growing varieties, or vines handicapped by unusually poor soil, require a third year before they can produce a cane suitable to serve as a permanent trunk.

3. TRAINING SYSTEMS FOR VINIFERA VARIETIES

Since the growing habits of *vinifera* varieties and native hybrids differ somewhat, they are trained according to different systems.

Head training, which assumes the use of spur pruning, is the most popular form for the wine varieties grown in California. In the fullness of the season's growth a vineyard trained this way looks like a well-cultivated orchard of miniature trees or shrubs. A close-up of a single vine shows that it has a straight, unsupported trunk, from the head of which the shoots emerge with their burden of fruit and foliage, usually drooping umbrella fashion. The better to understand this form of training, let us examine one of these vines after the fruit has been gathered and the leaves have fallen. It has a trunk, usually 15 to 24 inches high, from the top (or "head") of which branch several short arms; the canes (now shorn of fruit and foliage) originate from spurs one of which was left on the end of each arm the previous spring. At the subsequent pruning one cane is left on each arm, and that one cane is pruned back to a new spur of two or three eyes.

With head training, the vine requires no supporting trellis. Since there is no trellis, cross-cultivation is possible. Also, it is easy to train and prune. Its defects are that very vigorous varities frequently fail to set their fruit well when pruned this way and bear unevenly, and that its use is limited to those varieties whose basal eyes are fruitful. Further, when applied to small-clustered varieties (such fine varieties as Pinot Noir, Riesling, and Chardonnay, for example), a light crop is the result. However, when it is applied to the typical heavy-bearing varieties of California varieties [1] it is entirely satisfactory.

The first step in forming a head-trained vine is to cut the trunk to the desired height, tie the top of this young trunk

[1] The suitable form of training is mentioned in connection with the description of each variety in Chapter XII.

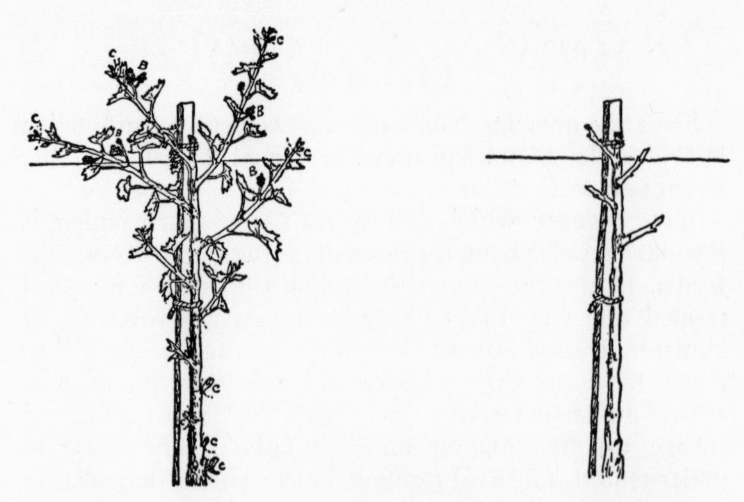

FIG. 16 (*left*). *Forming a head-trained vine. Permanent trunk was formed by cutting off cane which developed during the vine's first season and allowing shoots to develop from the top four bud-eyes only, suppressing all those lower down on the trunk.*

FIG. 17 (*right*). *Forming a head-trained vine. The four canes which developed from the top buds are pruned back to spurs of two buds each. Each bud will develop a shoot during the next growing season, and each shoot will carry 2 to 4 bunches of fruit.*

securely to its stake, and make a second and looser tie about half-way between the top and the ground. In the spring, when the buds begin to push, all buds but the top four or so should be removed. Thus several strong shoots will develop near the top of the trunk (see Fig. 16). Adventitious shoots below these should be rubbed off, and blossoms appearing on the shoots should be nipped.

During the subsequent dormant season all four of these shoots are cut back to spurs of two eyes each (see Fig. 17). Each of these 8 eyes will produce a fruitful shoot, and the shoots will bear a small crop. During the *next* dormant season, the four or five canes that are most symmetrically placed are pruned back to spurs of two or three eyes each, and the rest are entirely eliminated. From then on, the number of spurs left is a matter of individual judgment. With time, as shown in the photographs in Plate VI, a vigorous vine may be able to carry a dozen spurs of 2 to 4 buds each.

The management of a head-trained vine calls for some discretion. During the first years, the spurs to be left on the young trunk to form the arms must be chosen with care, and spaced as evenly as possible about the head in order to give it good form and prevent the crowding of fruit. Later, those canes which are to be cut back and left as spurs must be so chosen as not to lengthen the arms unnecessarily. And when an arm gets too long, one must watch for a suitably placed water sprout located back on the arm, and, during the following dormant season, prune this to a replacement spur so that the part of the arm lying beyond it may ultimately be pruned off and dispensed with. Finally, judgment is needed in deciding how many buds to leave on a given spur. A spur on a weak arm may be shortened, so as to encourage vigorous growth; that on a strong arm may be left an eye or so longer so that it may bear more fruit but make less growth. Thus the vigor and fruiting capacity of the various arms (which may vary from three to eight in accordance with the vine's capacity) are kept in balance.

Cane training is the form that allows the use of mixed (i.e., cane-and-spur) pruning, and in California it usually involves a two-wire trellis, the bottom wire being about 28 inches from the ground and the top wire about 16 inches above this. The bottom wire should be No. 10, and the top wire No. 12. The trunk extends to the bottom wire. The vine is so pruned as to leave two (four, if the vine is vigorous) canes, each with its spur. A cane and a spur are left on each arm. The canes, with 8 to 12 eyes apiece, are tied to the lower wire, to the right and left of the head of the vine. The canes produce the crop, the young shoots being tied to the top wire as soon as they make enough growth to reach it.

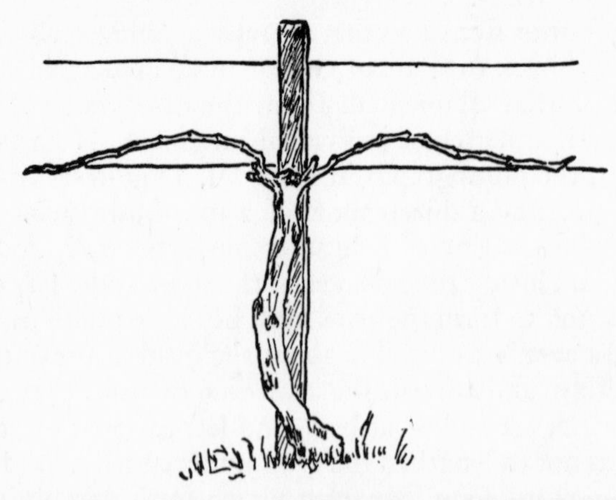

FIG. 18. *Cane Training. Preliminary training during the vine's first two years is identical with that for developing a head-trained vine. But at pruning time two canes and two spurs are left instead of four spurs.*

Cane training is necessary for varieties whose basal eyes are unfruitful. It should be used for all varieties whose grapes are small and develop in small bunches, if good crops are to be had. Since it allows a proportionately greater leaf area to develop during the early part of the season, it improves the health and growing capacity of the vine instead of diminishing it, as head training tends to do. It distributes the fruit along a cane instead of causing it to be bunched closely about the head of a vine. And finally, it is better for vigorous varieties that do not set their fruit well when trained to the head form.

Its principal disadvantage is that too many eyes may be left and the vine may thus be forced to overbear; and this means not only an exhausted vine but inferior fruit and irregular bearing until the vine recovers. But the vine-dresser who is aware of this danger and prunes accordingly has no reason to worry. The other disadvantage lies in the fact that cane training requires a trellis. Those who prefer cane training, but do not care to invest in a trellis, may get around the difficulty by adopting one of the several cane-pruning

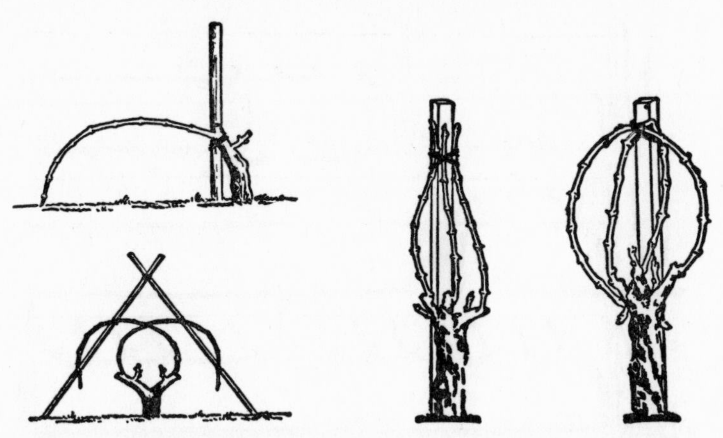

FIG. 19. *Variants of cane training which do not require a wire trellis.*

systems which are shown on this page and which are customary in certain European regions.

The pruning of cane-trained vines requires more skill and judgment than the pruning of head-trained vines. Since comparatively few canes are retained, they must be carefully chosen – well matured and well placed in relation to the vine's permanent form. A cane-trained vine that is badly pruned for several successive seasons quickly becomes unmanageable and is not easily brought back into good shape.

Cordon training is used a good deal in California and in Europe for the table varieties of *vinifera.* In essence, it is a hydra-headed head-pruned vine with a long, horizontal trunk instead of a short, vertical one. That is, the spur-bearing arms are not bunched about the top of the trunk but are spread along its side. Sometimes, with ultra-vigorous vines, each arm along the cordon is pruned not to a spur but to cane-and-spur combinations. (See Fig. 20.) Cordon-trained vines require some skill to establish. But, when applied to appropriate varieties, cordon training produces huge crops.

Training for Winter Protection. When *vinifera* are grown in a region subject to severe winters, they are frequently

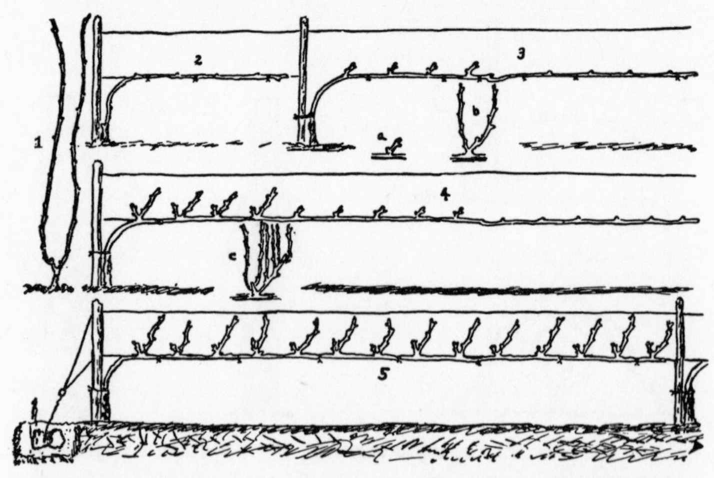

Fig. 20. *Forming a cordon-trained vine. The sequence of steps extending over a number of years* (1 *through* 5) *is self-evident. Close-ups indicate:* (a) *2-bud spurs left at beginning of third season;* (b) *development of the 2-bud spurs;* (c) *development of short cane-and-spurs.*

killed to the ground, and hence rarely produce crops, unless they are protected. For this purpose, conventional training methods must be modified. In climates otherwise tolerable to *vinifera* grapes but with cold winters – such as parts of New Mexico, Arizona, and Utah, and a good many parts of central and eastern Europe – a favorite method is to train the vines to the head form but with an extremely short trunk – say, four or five inches. In the autumn, the canes are tied over the vine's head, like a little girl's hair when she takes a bath, and soil is then banked up over the vine with plow and spade. Pruning takes place after disinterment in the spring. Another method, called *en chaintre* by the French, has been used with success by one *vinifera*-grower in Utah. He trains the trunk very low to the ground, with spurs emerging here and there along its length. In the autumn this low horizontal trunk is completely buried. In the spring it is disinterred, lifted, and held clear of the ground with a forked stick or by tying it to a stake, or by tying it to the bottom wire of a low trellis. The third method is a modification of cane training. The trunk is left very low, and with the approach of winter the canes to be used for canes and spurs are bent gently to the ground and buried. In the spring, these are disinterred and pruned, the fruiting canes being tied up to a trellis wire. The bunches of grapes are thus kept fairly clear of the ground – especially if the relatively unfruitful lower buds are rubbed off – and are thus less subject to mold and to spattering by mud.

4. TRAINING SYSTEMS FOR THE HYBRIDS

All good training systems for native hybrids involve the use of the spur-and-cane idea. Some of the French hybrids are adapted to head training with spur pruning, as described for the *vinifera* grapes of California.

Kniffin Training. This is the most popular training system in the eastern regions. The photograph in Plate VIII shows a young vine trained to the four-cane Kniffin form, just after it has been pruned. The vine consists of a trunk about four feet long, to reach to the top wire of a four-foot trellis. Near the head may be seen two canes and two spurs – the canes to produce the season's crop, and the spurs to produce the renewal wood for the subsequent season. At a point on the trunk near the lower wire, two canes with their spurs are also left. These lower canes are pruned somewhat shorter than the top canes, because the growth of their shoots tends to be less vigorous. Shoots from the eyes of the four canes droop downward (a normal way of growth for many of the native hybrids) forming a mantle of foliage for the fruit.

For vigorous vines, it is customary to leave 40 to 60 fruitful eyes per vine – say, 12 to 16 eyes each for the two top canes, and 8 to 12 eyes each for the two lower canes. Figuring two bunches for each growing shoot, this gives a crop of around 80 to 120 bunches per vine, which is about as much as the average vine is prepared to ripen. Weak-growing vines must be pruned more severely.

The Kniffin system distributes the grapes evenly over the

entire trellis — an important advantage in humid climates — and keeps much of the crop fairly high. Its only important disadvantage is that the top canes tend to rob the lower canes of their sap and thus to discourage their growth.

Six-cane Kniffin training is an obvious modification of the standard four-cane Kniffin sytsem. It requires a three-wire trellis (one wire for each pair of canes), and the trunk must be trained correspondingly high — five or six feet. The advantage claimed for this form is that, since it allows the vine to expand vertically, the length of the individual canes may be shortened somewhat without any loss of yield; and since the canes are shorter, the vines may be placed closer in the row and there is a larger aggregate yield. On rich soils, this is undoubtedly true.

Umbrella Kniffin training is a modification of the basic Kniffin form in the direction of fewer, rather than more, canes. It is increasingly used in commercial vineyards throughout the East. The lower pair of canes is dispensed with entirely, and the vine is annually pruned to two long canes (with accompanying spurs), which are draped over the top wire, brought to the lower wire, and there tied. Since the two canes are proportionately longer, the elimination of the two lower canes of the Kniffin system means no great loss of crop. In the case of very vigorous varieties, four such canes may be left.

Munson System. T. V. Munson worked out a modification of the Kniffin idea which has special advantages for hot regions, such as Texas and Florida, and for humid regions. It is also used in California in a few vineyards for choice table grapes. The difference is not so much in the training of the vine as in the trellis. A Munson trellis looks for all the world like a row of small telephone poles, each with its cross-arm. Three wires are required for the Munson trellis. Two are fixed to the outer ends of the cross-arms. The third runs through the posts at a point three or four inches below the cross-arms. The trunk of the vine is brought up to the middle (and slightly lower) wire, and there tied. From this head run canes, one in each direction from the trunk, along

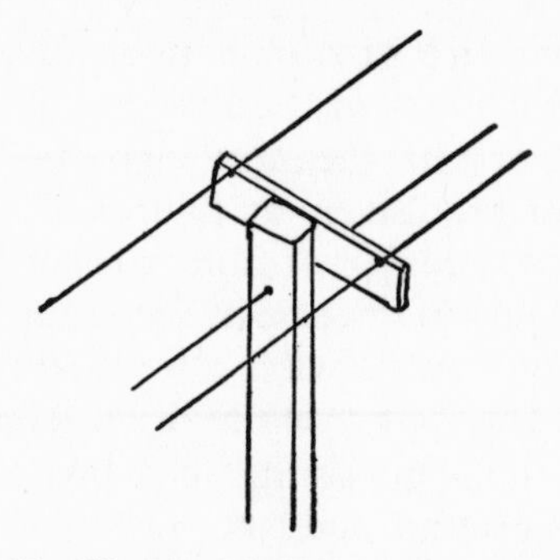

FIG. 21. *Trellis for Munson system of training.*

the middle wire. Renewal spurs are left at the head, of course. As the fruit-bearing shoots develop from the eyes of these canes, they grow out and over the two side wires, and then, following the custom of our native vines, droop. Thus shoots and foliage form a tent for the developing grape clusters, which hang sheltered within it and free of entanglement with shoots, leaves, and tendrils. A sufficiently vigorous vine will support four fruiting canes, two running each way from the head along the center wire.

This arrangement shades the grapes well in hot regions, and gives them good ventilation.

The main features of its construction are clear enough: good stout end-posts with strong cross-arms, preferably bolted and braced; lighter intermediate posts with two-by-four cross-arms, rigidly secured so that they won't wobble with age; the center wire run through holes bored in the posts, and the two outer wires secured by staples. Munson recommended four feet as the most convenient height.

High Renewal Training. This is the name applied in the East to a system that is practically identical with cane training as it is described for *vinifera* grapes on page 117. It differs from the Kniffin systems in that it employs a three-wire trellis, and the grape-bearing shoots, instead of being allowed to droop, grow vertically and are either tied to or interlaced between the wires as they grow. It is ideal for the weaker-growing varieties, such as the Delaware, and especially when they are grown in poor soils like those encountered on the shores of the Finger Lakes. But since it is entirely satisfactory for the more vigorous varieties as well, we use it almost exclusively at Boordy Vineyard, regulating the length and number of canes to suit the requirements of each variety, and in the case of vigorous varieties looping the canes over the middle wire. It is also the method largely used in Europe for the French hybrids. This form of training requires a trellis consisting of a lower wire 28 to 36 inches from the ground, and two wires spaced about 14 and 28 inches above the bottom wire. The trunk extends to the bottom wire only. The vine shown in the two photographs in Plate V is trained in this way.

Cordon training. For certain ultra-vigorous hybrid varieties we have found cordon training, as described on page 119, to be highly satisfactory.

5. PRUNING PRACTICE

When to Prune. One may begin to prune any time during the dormant period, except during extreme cold when the canes are very brittle. The pruning of large vineyards thus often extends over the entire period from late November until April, whenever outdoor work is feasible. But in cold or frosty locations there are reasons why too early pruning is undesirable. For it sometimes happens that canes are partly killed back during severe freezes, or that entire canes are thus destroyed. By postponing the pruning until spring, it is possible to see just what canes have been affected by the cold, and to what extent, before deciding which canes to leave for crop and spurs and which to remove.

We know that late pruning, even though it is delayed until the sap is running, does not injure vines. In frosty localities it is actually an advantage thus to delay pruning, even until the first buds have begun to push forth their shoots. Most varieties do not push all buds simultaneously but push their end ones first. Thus, the outermost eyes of a cane may already have shoots several inches long before those farther back toward the trunk have begun to open. Several weeks often intervene between the opening of the first buds and the opening of the last. So the cautious vine-dresser sometimes leaves his canes rather longer than he expects them ultimately to be. If frost comes, only the advanced shoots from the end buds are affected; and he may then prune the canes back to their "normal" length. The eyes on which he counts for his crop have escaped the frost altogether.

Yet it is sometimes inconvenient to delay all pruning until very late, and many growers compromise. They make a preliminary pruning during late autumn and early spring, eliminating most of the wood then but leaving enough sound canes to provide insurance against misfortune. Then later in the spring the vines are gone over once again for their final pruning.

How to Prune. The traditional European pruning tool is the *serpette*, a knife with a hooked blade. It makes a clean, sharp cut and allows the work to proceed swiftly. But unless carefully used, it injures a good many canes; and pruning tends more and more to be done with shears. These are satisfactory, though the single-bladed type sometimes crushes the end of the cane as it cuts. To avoid this, prune so that the blade of the shears is closest to the vine; then any crushing that is done hurts that part of the cane which is discarded.

For cutting the heavier wood, when that is necessary, some use two-handed pruning shears. But here the danger of crushing the wood, and thus leaving a wound into which rot can enter, is more serious. The alternative is a small fine-toothed saw.

In all trellis systems, the canes must be tied. If the cane is rather long, it may be twisted once gently around the wire and tied at the outer end only. If it is short and stiff, two ties may be necessary. But except at the outer ends of canes, where a good tight half-hitch is the proper knot, *the tie must be loose;* in the course of a season, especially with young vines, trunk and canes thicken in astonishing fashion; and unless plenty of room is left, there is either a broken tie or a girdled vine. There is a great choice of tying material.

Around Keuka Lake pliable willow is used, even as used in the Roman vineyards of Cato's time. Binder twine is cheap and satisfactory, as is very fine wire.

Green Pruning. The pruning so far described is all to be classed as dormant pruning, even though it takes place late in the spring. The term green pruning is used to describe several supplementary treatments, given while the vine is in active growth: *disbudding*, *pinching*, *girdling*.

During spring and early summer, shoots that serve no useful purpose appear from adventitious buds scattered over the older parts of the vine. They appear commonly on ungrafted vines, near the crown – that is, from the region where the trunk merges with the roots – and from the head. They waste the vine's energy, and they should be suppressed. The best way to do it is to rub or break them off while they are small. If this is done systematically, by making several complete rounds of the vineyard early in the year, especially when vines are young, later trouble is avoided.

Grafted vines require another form of disbudding, called suckering. This is the removal of any young shoots which may start up from the stock and which, by diverting sap, seriously weaken the scion. These must be regularly suppressed during the first few years after planting, or the scion dies. Young roots sent down from the scion, in an effort of the scion to bypass the rootstock, must also be suppressed during the graft's early years.

There is no question about the value of disbudding. In contrast, *pinching* – the pinching off of the growing tips of green shoots – has been the subject of much controversy. This arises from the fact that the removal of parts of shoots involves the suppression of leaves, the factories in which the vine's food is elaborated, and thus tends to diminish the vine's capacity for growth. If all of the vine's leaves are removed as rapidly as they appear, the vine dies of starvation; if some of the leaves are removed, in just that degree it is deprived of nourishment.

Pinching is used chiefly to hold back the growth of the more vigorous shoots in order to direct and concentrate growth in the less vigorous. When the growing tip of a shoot is removed (it is literally pinched off) the effect is to halt that shoot's growth temporarily. Before long it resumes growth by putting forth laterals. But in the meantime the sap is redirected into the less vigorous shoots and they are enabled to catch up. As we have seen, end shoots often make excessive growth. By pinching them back two or three leaves beyond the shoot's last inflorescence, the other shoots are given a better chance. It is sometimes advisable to pinch the growing tips of ultra-vigorous shoots several times.

Pinching is also helpful in directing the growth from the fruit-bearing shoots to those shoots which are destined to serve as next year's fruiting canes. It is useful also in windy regions, for, by slowing down the shoot's growth, it lets it gain strength near the base and thus strengthens it against wind damage. In hot regions, pinching, since it promotes the growth of laterals that shade the grapes, is also practiced to some extent.

Girdling differs from other forms of summer pruning in that it is practiced on the wood rather than the green shoots. It consists in the removal of a ring of bark from a cane, and its effect is to concentrate the nutrient materials elaborated in the leaves in those parts of the vine which lie out beyond the incision; if the girdling ring is complete, these materials cannot flow back, as they would normally, to the rest of the vine. Girdling weakens the vine permanently if it is done on any permanent part of the vine. If it is confined to a cane that will be eliminated at the end of the season anyway, the injury is less and the beneficial effects sometimes outweigh it. The practical effects of girdling are three: improvement of the "set" of berries at blossoming time; increase of the size of berries; and the hastening of maturity. Since a girdling incision heals rapidly, its beneficial effect is short-lived and the time for girdling is controlled by the specific effect desired. If an improved set is wanted, the girdling must be done during or just before blossoming. If larger berries are wanted, it should be done *after* the berries are set and during the time when the grapes are making their greatest growth. If early ripening is the end in view, the girdling should be done when the grapes are about half-grown, and the wound must be kept open.

Girdling is helpful chiefly to growers of choice table grapes. For the wine-grower, it is useful only to improve the set – and then, only the set of certain varieties (the Malbec particularly) that are known to respond well to the treatment. It is a delicate operation, and if done at all should be done with extreme caution, so as to remove all of the bark in a complete ring, yet not cut the wood itself. It is commonly done with a special girdling knife, in the handle of which two thin parallel blades are set about $\frac{1}{8}''$ apart. The edges of these blades are concave, in order to facilitate the operation of "rolling" the blades around the part to be girdled. In Europe, a small instrument resembling a pair of pliers, but with double blades where the teeth of the pliers would ordinarily be, is commonly used for making the incision.

Chapter IX

VINEYARD MANAGEMENT

Let us assume that the vineyard has been well prepared and planted following the suggestions in Chapter VII. During the first season the young vines receive their preliminary training as described in Chapter VIII, which is to say that one strong shoot is allowed to grow from the base of each vine and is tied at intervals during the season to the stake or lath placed beside it.

During this first season the new vineyard must be kept thoroughly cultivated to hold down competition from weeds and encourage growth. In dry spells it should have water, as young vines with their limited root systems are more sensitive to drought than established vineyards. Cultivation may be stopped toward the end of August to encourage the ripening of the shoots (now becoming woody) which are to become the trunks of the new vines. At this time the vines may be hilled lightly. With the last cultivation a light sowing of domestic rye grass or rye may be laid down in the alleys to serve as a cover crop, in case the vineyard is on a slope and subject to erosion.

The permanent trellis is put in during the following first dormant season.

The question now is how to maintain the new vineyard and how to do this with the least outlay of time and effort. A vineyard is not a garden of rare and fragile plants that must be petted and coddled. But it will not thrive on neglect.

1. A CALENDAR OF LABORS

The calendar of vineyard labors begins in late winter or early spring while the vines are still dormant, with *pruning* according to one of the methods already described. For pruning the needed equipment is a pair of sharp pruning shears, coarse twine (binder twine will do), and a knife. A carpenter's apron is convenient for holding these, and the work of tying the newly pruned canes is speeded up by the use of a finger knife, which fits on one of the fingers like a ring and has a little hook blade. The pruning proceeds rapidly. When it is finished the prunings are gathered in heaps in the centers of the alleys and hauled out. A heavy-duty rotary mower or chopper, if available, can chew them up where they lie, to be worked into the soil with cultivation.

Tilling, the next step, should be done before the buds begin to open, though not when the soil is too wet and heavy; for a wet and heavy soil cakes and clods. The traditional practice is to plow, turning the soil away from the vines in the spring and toward them in the fall.[1] But except in special circumstances plowing is now considered bad practice since a vine's feeder roots lie near the surface and plowing damages them, and since the sole purpose of vineyard cultivation is to hold down weed growth. Modern tillage tools have made plowing unnecessary.

In a large vineyard, the first spring cultivation is done with one or another of the conventional heavy-duty tractor-drawn tools, usually a disk harrow set fairly shallow, which chops up weed growth or cover crop and leaves a trashy mulch partly worked into the soil. If cover is heavy, a spring tooth harrow may be used to rip it up first, followed with disks. A variant is the big tractor-drawn rotary tiller, also set shallow.

A little later, but before too much vine growth, the ridge of soil beneath the trellis row is pulled away by means of a *grape hoe* (usually hydraulically operated these days) which is fitted to the side of the tractor. This moves in and out between the vine trunks, leveling the soil under the trellis and removing all weeds.

For working a small home vineyard the equivalents are

[1] The two plowings are called *réchaussage* and *déchaussage* in France.

obvious: a garden tractor equipped with disks or cultivating teeth or better yet a rotary garden tiller. A word of caution, though. Many of these power garden tools are built down to a price and are hardly more than toys intended to catch the eye of innocent suburbanites. It is a waste of time to buy anything but a durable machine with a sturdy engine and cultivating parts, and this goes double if the vines are planted in stony or heavy-textured soil.

The best garden tractors or tillers for use in a small vineyard are those with offset or swiveled handles, allowing cultivation close to the vine row. The closer this cultivation, the less hand work. Whatever ridge of soil remains beneath the trellis must be removed by hand, the best tool for this purpose being a heavy, short-handled, broad-bladed hoe of the kind known in France as a *pioche*.

This first cultivation ought to be done before vine growth has become so abundant as to deny a close approach to the row, and before new weed growth has gone much beyond the seedling stage—also to keep spring growth of cover crop from jumping out of control. Subsequent to this principal tillage, the work of cultivation consists of occasional use of disks or harrow to suppress the later crops of weeds. Normally one of these bouts takes place right after blossoming, when the berries have set, in mid-June; another in early July; a third in late July or early August. It is the amount of weed growth that controls this, more cultivation being required in a wet season than in a dry.

If considerable weed growth occurs in the strip beneath the trellis, especially where wild morning glory is prevalent, a round of mid-season grape-hoeing may be necessary—a throw-to of soil to smother followed ten days later by a take-out. The equivalent in a home vineyard is a bout of hand-hoeing.

Not long after the *véraison*—the turning point in the vine's growth in early August—the vineyard is given its final cultivation, a ridge of soil is thrown to the vine row with the grape hoe, and cover crop may be planted. At the *véraison* the green bunches begin to change color and begin to ripen, and the shoot growth likewise begins to ripen into wood, this phenomenon beginning at the base of the canes and extending outward. Most growers at this point are content to let the weeds take over for the rest of the season, and the facetious call this weed growth their cover crop. As a matter of fact it is, and it provides many of the functions of a more sophisticated cover crop. More about this later.

The time from the *véraison* onward to the vintage is the leisure period for the wine-grower. The wait is long or short depending on the variety of grape. His principal occupation is worrying about deficiency or excess of rainfall, the risks of hail damage which can wipe out a fine crop in a minute or two, the onslaught of vine diseases in varieties susceptible to them, bird damage, little-boy damage, and assorted other hazards. The means of judging ripeness for wine are discussed on pages 13 and 90. In California the important thing is to pick soon enough, while the fruit still contains sufficient acidity and before it develops too much sugar. In most other parts of the country and particularly in those parts where there may be much end-of-season rainfall and humidity, the important thing is to curb one's impatience and wait long enough to gain maximum sugar content and minimum acidity. The temptation to pick too early, under such conditions, is strong. It must be resisted.

The actual harvesting should be done on a bright, breezy day so that the grapes are not wet; and preferably in cool weather so that the grapes will not come too hot to the fermenter. The picking is done with a knife or with grape snips. The grapes ought to be handled carefully, so that they are not bruised; and if conditions have caused some rot to develop, the time to cut it out is while picking, not later. Containers for picking should be baskets or plastic or wooden "lugs," not metal.

Once picked, the grapes should be kept in the shade; and to reap the utmost reward in quality of wine the wine-grower will see that they are brought to his crusher with no great waste of time. Careful wine-makers like to see the grapes all crushed by the morning after they have been picked. If held overnight, they should be covered with cheesecloth to keep down the yellow jackets and fruit flies.

2. ALTERNATIVES TO TILLING

The routine already described could properly be called *modified clean culture*, since it combines clean culture from early spring until late midsummer with ground cover from then on through the winter to spring again. In this context, "clean culture" is relative. A trashy and highly absorbent surface mulch which leaves a few weeds is preferable in most cases to the perfect cleanliness obtained by turning the soil deeply and all the way over with a plow. Yet in areas subject to prolonged drought plowing can be an advantage. By damaging surface feeding roots it drives the root system deeper and so makes the vine's feeding system less immediately responsive to dry spells.

Then there is *clean culture* pure and simple, the object of which is to keep a mulch of loose surface soil all the year round. A leading exponent of this system was E. Maroger, a grape grower of southern France who achieved remarkable yields by a shallow cultivation once a week all the year round, with a view to absorbing maximum moisture and by keeping the surface layer well aerated encouraging the nitrogen-fixing bacteria.

Weed Killers. It must be clear from the preceding pages that the most difficult and expensive part of cultivation is that of keeping the narrow strip beneath the trellis rows free of weed growth. This requires the slow and expensive and sometimes damaging work of machine grape-hoeing, and may require a good deal of hand hoeing besides. In principle the weeding of the under-trellis strips by means of chemical sprays offers great advantages, and the vineyard use of chemical weed killers has been under study for decades.

The problem of course is to find a chemical that can distinguish between weeds and grapevines, thus doing away with the weeds but not killing or injuring the vines. And here we move into tricky territory. The most widely prevalent weed killers, 2, 4-D and 2, 4, 5-T, are deadly for grapevines. They must not be used in vineyards or even near them or in any equipment that is to be used in the vineyard.

However, more selective toxicants have been developed and are now widely used in commercial vineyards. As recently as 1957, better than 95 per cent of the New York commercial vineyards were mechanically cultivated throughout. Today in at least half of them weeds are controlled in the trellis row by means of spraying, with substantial saving of labor cost. The materials used are of two kinds: the so-called selective soil sterilants known as Karmex and Telvar and the dinitro contact sprays. And now the use of weed sprays is extending from the under-trellis strip to the entire vineyard floor—doing away with mechanical cultivation completely.

Their use remains extremely tricky. Application must be closely calibrated lest too much or too little be applied. Timing is important and of course is different from one region to another. Whole families of weeds, including bindweed, milkweed, chicory, Canada thistle, and many other first-rate nuisances are unaffected or only partly controlled. Even though vines may suffer no apparent damage, the actual effect of these compounds on the life of the vine is inadequately known. What is known is that they cannot be used on young vines (three years old or younger) without apparent damage, which suggests that the materials are hardly likely to do older vines any good.

The net of the matter is this: that amateur wine-growers will be wise to leave the weed killers entirely alone and under no circumstances to use them on immature vines, and that commercial growers should use them only in close consultation and cooperation with specialists in their area.[1]

[1] See "Chemical Control of Weeds in New York Vineyards" by Nelson Shaulis and T. D. Jordan. Cornell Ext. Bul. No. 1026.

Sod Mulch. The opposite of the barrenness of control by weed killers is the maintenance of a vineyard under permanent sod, keeping it closely mowed. Traditional opinion is against this. The vine responds remarkably to tillage—so remarkably that where cultivation is possible it has invariably been practiced. This is the case even where elaborate terraces must be built to protect the bare soil from erosion, as along the Moselle, on the steep slopes of Lake Geneva, on the mountainsides of the upper Douro in Portugal. Yet it seems possible that sod culture of vineyards has been too thoroughly condemned. Certainly the question is worthy of review.

Against it is the undeniable competition between sod and vines for available soil moisture and plant nutrients. In areas of low rainfall the sod can rob a vineyard of moisture to the point where it shows symptoms of drought; yet a neighboring block of clean-cultivated vines may continue to flourish.

Favoring it is the capacity of sod cover to increase water absorption. When rainfall comes in a downpour, clean-cultivated soil can often not absorb it all. A heavy rainfall on clean-cultivated soil causes erosion; both sheet erosion, which is not always noticeable, and the more spectacular gully erosion. Sod hinders run-off and gives the ground more time to absorb the water. These are important virtues for vineyards located on slopes or on light-textured soils.

There is another reason for reconsidering sod cover for vineyards. Sod management used to be difficult between vineyard rows. Sidebar mowers couldn't get in, and scything is slow and laborious. Tractor-driven rotary mowers have changed that. It is no trouble with one of these to hold vineyard sod under close control with regular cuttings —all of it, that is, except that bothersome strip directly beneath the row. But the solution of this detail is not beyond human ingenuity.

It may be now that its advantages outweigh its disadvantages and that the whole question is worth experimental review "in the field," as the farm scientists would say.

3. SOIL TYPES

This is as good a place as any to correct some common misconceptions about the importance of soil types in producing good wines. An impression prevails that the nature of the soil is all-important to wine quality. It is true that in every important grape-growing area certain types of soil and pieces of land are preferred over others. There are soil conditions, moreover, that are hostile to grapevines: waterlogged soil, very shallow soil over rock or an impervious subsoil, alkali soils, and others that are actually toxic. But when that is said, the fact stands that the grapevine is one of the most broadly tolerant of plants so far as soil is concerned. For vines the so-called *p*H range is extremely broad, extending from acid conditions to the soils running as high as 70 per cent of calcium carbonate in the chalk of the French champagne district.[1] Going over the catalogue of the world's good wines, one finds that some are grown on sand, some on gravelly sands, some on loam, some on fairly heavy clays, some on deep soils, and some on shallow. The French Côte d'Or is on the alkaline side; soil in the neighboring Beaujolais is granitic and hence acid; Champagne and the upper Moselle, chalky; the German Palatinate, gravelly. Winkler remarks that the Schloss Johannisberg vineyards of the Rheingau are on red shale, whereas the nearby and equally celebrated Steinberg vineyards are on clay loam. In all the other districts from Italy to California there is an equal diversity.

Then why so much emphasis on soil types, especially in the European wine literature? Custom, tradition, local familiarity with the *handling* of certain types of soil, the fact that some soils in cold climates are "warmer" than than others and so induce earlier ripening—all these are contributing factors. Then there is the natural and unbreakable monopoly to be had by associating a certain tightly

[1] But this high tolerance for lime is a characteristic of the *vinifera* varieties only, growing on their own roots. The American species and hybrids containing them are subject to chlorosis when lime content goes beyond about 20 per cent.

held piece of real estate—Romanée Conti, for instance, or the Clos de Vougeot—with a given wine name.

Yet as a determinant of quality the type of soil runs a poor third to climate and the variety of grape being grown. Let the climatic conditions be satisfactory and the grapes be of a superior variety and suitable for that climate, and the grower need not worry too much about his soil type, provided it be reasonably well drained and of middling fertility.

4. MAINTAINING FERTILITY

With that misconception out of the way, let us return to our calendar of labors. Literally followed, it will be satisfactory in most circumstances. Yet soon or late most growers feel the need for a better understanding of the process of growth and nourishment and the impulse to introduce variations into their routine. They wish also to guard against the impoverishment of the soil.

What is meant by the impoverishment of the soil? The vine's food, like that of other green plants, consists of water, which is absorbed mainly from the ground, an assortment of minerals which are absorbed with the water as soluble salts, and oxygen and carbon dioxide, which are taken by the leaves from the air. Normally there are only three minerals whose presence or lack is of real concern to the grape-grower. These are nitrogen, phosphorus, and potassium. Nutritional trouble in the vineyard—"impoverishment of the soil"—is usually due to the deficiency or the depletion of one of them.

Nitrogen directly affects the vine's vigor. Give a vine much nitrogen, and it grows furiously. Leaves are large, shoots are large but relatively soft, and the shoots grow to abnormal length. But excessive nitrogen may act like a boomerang, because excessive vegetative vigor tends to diminish fruitfulness and the softer character of the growth may reduce winter hardiness and make the vines more liable to disease. The traditional prejudice against growing grapes in the rich deep alluvial soils of valley floors, rather than on the more meager soils of the slopes, boils down to a prejudice against too much nitrogen. Yet without an adequate (as distinct from excessive) supply of nitrogen there is an immediate falling off in both crop and vigor.

Potassium is required in larger quantities by the vine than by many other plants. Both shoots and fruit are rich in the salts of potassium, and potassium bitartrate (cream of tartar) is responsible for most of the acidity that gives wine its liveliness. A wine deficient in acid tastes flat. In the vineyard potassium has an intimate connection with the formation of carbohydrates in the leaves—of starch, that is, which in turn becomes the grape's sugar—so intimate that when there is a deficiency of potassium it is immediately shown in the sickliness of all green parts. A shortage of potassium is less frequent than a shortage of nitrogen. But a shortage when it occurs is only too evident.

Phosphorus seems to act as a sort of restraining influence on the exuberance caused by nitrogen. Wood matures more rapidly and completely; fruitfulness as distinct from vegetative vigor is improved; and sugar content seems to run higher. But most soils have a sufficient supply of available phosphorus. Neither in California nor in the East are supplements likely to have a measurable effect on grape production.

Other Elements. Another dozen elements[1] besides these three enter into the structure of the vine—carbon, hydrogen, and oxygen in large quantities from the air; the others in traces only, which average soils easily supply. But there are exceptions. Magnesium deficiency is not infrequent in the East; boron deficiency is encountered in some areas; zinc deficiency can be a real problem in certain California soils. In special situations an excess of one or another of these trace elements can also be toxic to grapevines. But here we move into complexities of soil science beyond the reach of the amateur and still full of mystery even to the

[1] Boron, calcium, carbon, copper, hydrogen, iron, magnesium, manganese, molybdenum, oxygen, sulphur, zinc.

specialists. The more soil is studied, the more baffling its physics and chemistry, and indeed its biology.

Organic Matter. An adequate supply of organic matter—humus—is important for many different reasons. It increases the soil's water-holding capacity and thus both reduces run-off and provides some insurance against drought. It gives the soil what farmers call good tilth. It keeps the soil "alive" in the sense that it is necessary to the multiplication of a wide range of microorganisms whose effect on the growth and health of the vine is obscure but undeniable. It slows down the leaching of nitrogen and potassium and helps to hold phosphorus in available form. A direct relation between humus content and the health and productivity of a vineyard has been shown many times. Constant cultivation tends to use it up, and the farsighted grape-grower does what he can to replace it.

5. PRACTICAL FERTILIZING

These remarks on soil and the nutrition of the grapevine point to some practical conclusions.

The first is that a new vineyard should be given a good start. Grapevines are not particularly demanding, but there is no point in handicapping them by bad location. Most of us who have had any experience of growing things can tell the difference between land of reasonable fertility and a barren patch. This impression can be confirmed by a soil test. The nearest county agent will arrange for one, and if there is a critical deficiency it will so indicate. He can also warn about any local peculiarities in the way of mineral deficiencies or toxic excesses. (But don't let him tell you to spread lime!)

Assuming a normal initial fertility, the wine-grower's interest is to maintain it. But this is nothing to be anxious or excited about. Remember that 99 per cent of the material of a grapevine's growth consists of water plus carbon dioxide taken out of the air. This means that the drafts on the soil's mineral supply are not heavy. Nothing is gained by an oversupply of the critical mineral elements, and indeed an oversupply is more likely to be harmful.

A sound rule is to do nothing out of the ordinary if the vines are healthy and vigorous and producing satisfactory crops of fruit. Shaulis puts it this way: "Vines without symptoms at the previous harvest will not grow or produce more as a result of mineral fertilization."

A corollary of this rule is to adopt a temperate routine of soil maintenance. This involves, first, some attention to the soil's organic content. An occasional dressing of manure is ideal, if manure is to be had, since manure provides not only organic material but a modest helping of nitrogen and potassium in available forms. In the absence of manure, any form of organic debris is useful—waste hay or straw, sawdust or wood chips, the pomace that is left over after the wine has been made—which may be spread between the rows during the dormant season and chopped in with disks. Two or three tons per acre of such material every other year are sufficient; and with this, 30 pounds per acre of actual nitrogen in one or another of its various forms should be spread and worked in to aid the process of breaking down the organic matter.

Another source of organic matter is of course a green cover crop, planted in late summer at the time of final cultivation and plowed in the following spring. A cover crop provides the equally important advantages of opening up compacted soil, improving the absorption of rainfall and reducing erosion. Domestic rye grass seeded at 8 pounds per acre is good for the purpose; so are ordinary rye and wheat, seeded at about 1½ bushels per acre. These should be disked in in the spring before they make excessive growth. Some wine-growers prefer to put down a cover crop only every other year in alternate rows, thus leaving a free alley for picking.

Now as to mineral supplements.

Of the critical elements, vines make the heaviest draft on *nitrogen.* Once a vineyard comes into bearing, an annual spring application of 40 to 60 pounds of *actual nitrogen* per acre is a sound rule. Nitrogen is available in various forms, of which ammonium nitrate is the best if the price is within reason. Ammonium nitrate provides 32 per cent of actual nitrogen, which is to say that a 100-pound bag provides 32 pounds. Thus an application of 60 pounds of actual nitrogen would require 185 pounds of this material. Urea (45 per cent actual nitrogen) is excellent too. In the absence of these, ammonium sulphate (21 per cent actual nitrogen) is always available.

Potassium does not have to be applied unless there are actual symptoms of potassium deficiency. The evidence is a browning or "burning" of the edges of the leaves from June on. If there is no such evidence, there will be no apparent response, in terms of improved crop and vigor, to additions of potassium. Even so, it is a safe practice to apply potassium annually at the rate of 100 pounds of actual potassium per acre. The best form is potassium sulphate (50 per cent actual potassium). Potassium should be applied in bands between the rows rather than broadcast. Potassium spread thinly tends to become "fixed" and hence unavailable to the vine roots. It may be applied whenever deficiency symptoms are observed, or at the end of the season.

Phosphate need not be applied regularly, although some growers do so (in late summer) to encourage a good stand of cover crop.

Lime should *not* ordinarily be applied to vineyards. An exception is an abnormally acid soil condition (with a *p*H below 5.4), which may be the cause of magnesium deficiency. In that case, ground limestone may be added at the rate of not more than 1 ton per acre, to bring up the *p*H. Excessive liming may actually induce potassium and manganese deficiency and is bad medicine for native varieties, for the French hybrids, and for *vinifera* varieties grown on hybrid rootstocks.

Complete Fertilizers. Because nitrogen and potassium fertilizers are most effective applied at different times and in different ways, separate applications in the proper amounts are to be preferred. But this is a counsel of perfection. Most amateurs and many commercial grape-growers prefer to simplify their routine by making a single application of what is called a "complete" fertilizer once a year.

A complete fertilizer is a formulation or mixture of all three critical elements—nitrogen, phosphorus, and potassium—in various proportions, plus other elements in some mixtures. The proportions of the *actual* quantities are required by law to be indicated on the bag, as for instance 10-10-10, 15-5-10, and so on. The first figure always indicates nitrogen; the second, phosphorus; and the third, potassium; and each figure indicates the proportion of the actual element *per hundred pounds.* Thus a 100-pound bag of 10-10-10 contains 10 actual pounds of each.

The proper quantity to apply is computed from the formula. For example, an application of 50 pounds per acre of *actual nitrogen* will require 500 pounds of a 10-10-10 fertilizer, which will also supply 50 pounds each of actual phosphorus and actual potassium.

But remember that nitrogen is the most consistently needed by grapevines; potassium is not quite so frequently needed; and phosphorus is least frequently needed. Therefore the grower who prefers a complete formula is wasting his money if he buys a formulation high in the latter two.

Garden shops and farm-supply houses offer a wide range of formulations. For the grape-grower who does not want to bother his head too much about precise dosages, a rule of thumb is to use a complete fertilizer carrying the formula 15-5-10, or as close to that as the dealer can supply. A complete fertilizer is always applied in the spring.

Chapter X

THE VINE'S AILMENTS

GRAPEVINES, like human beings, are subject to various ailments. Also like human beings, their susceptibility varies according to the innate resistance of the individual, general physical condition, and the presence or absence of causative factors.

Some people seem more susceptible to tuberculosis than others; likewise some grape varieties are more susceptible to the fungus diseases than others. Human beings are in general more liable to ailment if they live under conditions of undernourishment and overcrowding; and likewise vines are more liable to ailment if they are ill-nourished and badly tended. To offer one more analogy, nobody comes down with typhus if there are no lice around; and likewise grapevines do not suffer from Japanese-beetle injury if there are no Japanese beetles around. Grapevines, therefore, are "only human," and the wise grower disregards the two problems of prevention and first aid only at his peril. But this chapter is written on the assumption that the reader has only an incidental interest in entomology, which is the science of insects, or in mycology, which is the science concerned with the fungi. A wine-grower's interest in insects and fungi is confined to the purely practical business of seeing that they do not affect his vines adversely, and he seeks the simplest and most efficient means of attaining this end. The pages that follow are therefore boiled down to that minimum of information which the practical grower will require. Those who find their morbid curiosity aroused are advised to browse elsewhere, in the abundant literature of the subject.

Some of the vine's ailments have an animal (usually an insect) origin; some have a vegetable (fungus) origin. Let us lump the rest together as "other ailments." Further on, a *routine for disease control* will be provided. The reader interested in the how but not the why may jump to it. But, by way of introduction and background, some notes on the more troublesome ailments and the means of coping with them may be useful.

I. FUNGUS DISEASES

Three fungus diseases lie in wait to do great damage—two of them prevalent in the East but not in California, one widely prevalent but exceptionally serious in California.

Powdery Mildew. This one, known in France as *oïdium*, was first identified there, whither it had been carried on some immigrant American vines. The *vinifera* grapes are especially subject to it. The fungus attaches itself to the green parts of the vine (leaf, shoot, fruit) and multiplies rapidly, apparent as a grayish powder. Leaves lose color, wither, and drop off; thus the vine is starved. Tender shoots are badly weakened so that canes do not ripen properly into wood. Berries show grayish patches speckled with brown and presently break open, exposing the seeds. Powdery mildew has a moldy smell, so that wine made of fruit even partly affected has a characteristic odor that the French call *oïdiée*. Stems and pedicels are also affected and wither prematurely. The disease does not attack the fruit once the fruit has begun to turn color.

Varieties differ in their susceptibility. The Carignane may be ravaged by it, yet Malbec and Duriff will come through without too much damage. Nearly all cultivated vines are susceptible to it to some extent, some of the natives least so. One motive behind the production of new hybrids has been to obtain vines more resistant to oïdium than *vinifera*. But among these too the degree of resistance varies a great deal.

A simple and effective weapon against powdery mildew was discovered by Marés within a decade of its appearance in France. The same agent of control is used today. This is finely divided ("micronized") sulphur applied as a dust to all green parts of the vine. Mechanical dusters are used in large vineyards, but there are inexpensive hand dusters for small vineyards. In California the vineyards get from two to six sulphur dustings a season, depending upon the severity of the disease. The treatments are preventive rather than curative. The three most important applications are made (1) just before bloom, (2) just at bloom, and (3) a week or ten days later. But prudent growers prefer not to take chances and they go the whole program, which is six applications. The best time for dusting is in bright weather immediately after the dew has dried and before the heat of the day. Applied at high noon in a California summer, sulphur can do damage to fruit and foliage.

Where powdery mildew is a problem in the East, the spray material called Bordeaux mixture (consisting of either copper sulphate or basic copper plus lime) is normally used instead of sulphur dust, its copper being the antiseptic agent. The reason for this switch is that most of the native American grapes are subject to sulphur "burn" of the foliage. On them the sulphur cure, in other words, can be as bad as the disease. Some of the new organic fungicides such as Captan and Folpet give good protection. More about all this later on.

Downy Mildew (Peronospora). This is not to be confused with powdery mildew. It is no problem in California but flourishes in all the humid parts of the United States. In Europe it was the first of the accidentally imported American vine scourges, preceding both oïdium and phylloxera, and in most parts of Europe the vineyards must now be regularly protected against it. The *vinifera* are extremely susceptible. Our native varieties and the French hybrids are much less so and a few of them are practically immune.

Since this fungus requires much moisture it is worse in years and regions of high humidity, especially when there are sudden changes of temperature and much cloudy and showery weather. Regions blessed with constant breezes, serving to keep the vines dry, are much less troubled.

Like powdery mildew, downy mildew attacks all green parts; but it focuses on the leaves forming brownish spots that are apparent on the upper surface, and on the undersurface the downy whitish patches of summer spores that give the disease its name. The effects, if the leaves are badly attacked, are the usual effects of partial starvation: a slackening of growth, a failure to ripen grapes and wood, and a general sapping of the vine strength. Injury to the fruit of some varieties may wreck the whole crop: a withering of stems and berries which renders the fruit worthless.

There is a powerful agent for controlling downy mildew: copper compounds, traditionally applied as Bordeaux mixture. Because copper does reduce the vigor of grapevines to some extent, there has been a trend toward some of the new organic fungicides instead, chiefly Captan. But where infestation is likely to be serious, copper remains the best preventive.

Black Rot. This is unknown in dry viticultural California. But it can be disastrous in most parts of the country, and in Europe, unless precautions are taken. Black-rot infestation may be severe one year and negligible another, depending on climatic conditions, but the disease is always lurking around.

It appears in a wet spring as small spots on the young leaves, its spores having spent the winter on the dead wood and on the shriveled and hardened grapes left over from the previous season. From leaves the fungus passes to the fruit, and it is by direct injury to the fruit that black rot does its worst damage. Its presence is not usually apparent until the grapes have been thoroughly infected and begin to color prematurely. Very quickly after this appearance of color the berry begins to shrivel and to darken still further, and on its surface are to be noticed ugly black dots. Within a few days the grape has shriveled and mummified. The crop may be completely ruined, or only partly ruined; or if the attack is light the injury may be confined to isolated bunches or even single berries in the bunch. Once the grapes begin to ripen, black rot makes no further inroads.

Moisture, heat, and still air provide the most favorable

conditions. Against this, Bordeaux mixture is also the traditional preventive—that plus the destruction of infested prunings and mummified berries. The organic fungicide Ferbam is a specific against it but is not effective against the other fungus diseases. Thus if Ferbam is to be used against black rot other materials must be used against the other two, as many a grower has learned by bitter experience.

Other Fungus Diseases. Another intermittent troublemaker under conditions of high humidity is *anthracnose,* though it is not one of the principal menaces. A dormant lime-sulphur spray can help to prevent it, as does the old reliable Bordeaux mixture. *Dead arm,* which has been noted in parts of California but is a more serious matter in the East, is a fungus disease, sometimes confused with winter-killing, that is caused by the organism *Phomopsis viticola.* Infection appears as black spots on the canes, and eventually it may kill back canes, arms, even the trunk of the vine. It is controlled by prompt pruning and destruction of dead or infected canes and other parts of the vine, plus a preventive spray of the fungicide captan, which is a specific against this organism. Several forms of *root rot* can cause damage in California and in the South when certain fungi penetrate large roots broken or bruised during cultivation. Serious infestations are rare, and the only effective treatment is the drastic one of rooting out the affected vines and leaving the site fallow for several years. A number of different "rots" also affect the ripe fruit when rain and high humidity coincide with ripening. At this stage when the fruit is soft not much can be done except to pick the grapes as promptly as possible, cutting out the rotted areas of the bunch as one picks. One of these rots, *botrytis,* is the same as the so-called "noble rot" of the French Sauternes district.

2. ANIMAL DISEASES

Phylloxera. Some day someone is going to write a history of this native American plant louse and the dreadful things it has done to the vineyards of the world and the many ways it has affected the course of Western civilization. Pending that, a few paragraphs will have to do. The life cycle of this insect is extremely complex, but only in its root-feeding form does it cause direct and deadly damage. In this form it feeds on both small and large roots of grapevines. Rootlets injured by this little yellow louse are of course promptly replaced if the vine is vigorous; but the tuberosities that it forms on the larger roots become centers of decay which eventually cause the vine's death. Sometimes, as in the French vineyards in the nineteenth century, destruction comes with fearful swiftness almost like a forest fire, wrecks the economy of an entire region, and even causes wholesale human migrations.[1] Sometimes, as in parts of California, the vigor of the vines growing in soils the phylloxera does not much care for succeeds in postponing death for a long time.

There exists an aerial form which punctures the leaves of certain native species and their hybrids for egg-laying purposes, producing galls. This form may assist the spread of the disease somewhat. But it is not essential to the spread of phylloxera infection, as Europe's tragic phylloxera epidemic showed only too well. The aerial form is not common in Europe. Means of controlling the aerial form by means of systemic insecticides have been under study for some time.

As for the deadly root form of phylloxera, there is no successful means of eradication. Soil that is 60 per cent sand or better is not pleasing to it, a fact which has kept certain California counties free of it. Several American species and their hybrids are highly resistant to it and continue to thrive even in phylloxera-infested soil. The only really effective way of growing *vinifera* grapes in infested soil takes advantage of this high resistance of certain other species—this is the method of grafting on resistant root-

[1] The settlement of Algeria by displaced wine-growers from southern France, and hence much contemporary history, is directly related to the phylloxera epidemic.

stocks, now almost universal in vinifera-growing regions. Grafting is discussed in Chapter XI.

Grape-berry Moth. This is the small brownish moth that flutters out of vine rows at certain times during the growing season. The damage is caused not by the moth itself but by the larvae which live in the ripening fruit. The annual cycle is as follows: the moth emerges from the pupal state during the spring and lays eggs, which are hatched; these in turn pupate, and it is the larvae or caterpillars of the second brood that damage the grapes. The trick is to kill the caterpillars of the first brood by means of DDT or some other insecticide.

Japanese Beetle. This highly destructive beetle first appeared early in the present century in New Jersey and has spread relentlessly ever since. In the absence of natural enemies it multiplies in vast numbers, and it is particularly fond of the foliage of grapevines. In regions of new and heavy infestation, unprotected grapevines are completely denuded, with resultant weakening of the vine and failure of crop. The beetles over-winter in turfy land as grubs, and begin to emerge from the ground as beetles toward the end of June. July is the time of maximum activity. In August the feeding subsides as the beetles return to the ground to beget the next generation.

Yet with appropriate measures the beetle may be kept sufficiently under control. Successive light sprays of DDT or other contact insecticides such as methoxychlor or Sevin concentrated on the new top growth as it comes along will hold the beetle at bay.

Nor is the long-run situation so grim as it seemed to everyone a decade ago and may still seem to the grower who encounters the Japanese beetle for the first time in great numbers. It has begun to find a more modest place in our balance of nature. The discovery of the milky disease that kills the grubs was the first great breakthrough. Areas of heavy infestation were systematically inoculated with the spores of this disease, and these inoculated areas have enlarged themselves naturally with measurable results on the beetle population. Certain parasitic wasps that feed on the grubs are now well established after being deliberately introduced and have made further inroads. In the older Japanese beetle areas this once formidable destroyer is now being reduced to a manageable nuisance.

Leaf Hopper. There exist several species of this tiny white or yellowish flying or jumping pest, which is to be found in most parts of this country and frequently does considerable damage to grape foliage. For a time the infestation in the San Joaquin Valley of California was so serious that it threatened to wipe out some of the vineyards. This is a sucking insect that injures the vine by puncturing the leaf on its under side and thus reducing the leaf's capacity to elaborate the vine's food. Its presence in dangerous numbers may be detected by slightly shaking a vine on a still day in midsummer, when the leaf hoppers emerge in a cloud. The contact insecticides such as DDT and Sevin are highly effective against them. In some areas the development of immunity requires changes or alternation of pesticides.

Leaf Roller. The leaf rollers are moths whose larvae literally form rolls out of grape leaves while eating them. In serious infestations they can cause heavy defoliation. They can be a source of trouble in Eastern grape-growing areas and in California also. In the Northeast there are normally two generations a season, in California three. The contact insecticides DDT and Sevin control them.

Grape Flea Beetle. This tiny steel-blue beetle is found in many parts of the East, and in Southern California also. It hibernates under bark and in piles of dead leaves and trash, and emerges about the time the buds swell in the spring. It feeds on the buds and on tender young shoots. The beetles lay eggs on canes and bark, and the larvae hatched from these promptly ascend to the leaves and begin feasting. Control: vineyard sanitation and DDT applied as a very early spray when necessary.

Rose Chafer. This brownish beetle, one-third to one-half an inch long, sometimes causes trouble in vineyards by feeding on grape blossoms. It feeds on many plants besides grapevines, including, as its name suggests, roses. The usual contact sprays take care of it.

Erinose. The erineum mite, invisible to the naked eye, occasionally congregates on the under surfaces of leaves, producing a wart-like swelling or gall on the upper surface and a white felty concavity on the under side. It is easily confused with the whitish felty spots caused by downy mildew, and occasionally confused with the leaf galls of aerial phylloxera. Erinose apears to favor the leaves of some varieties more than others. It is easily controlled because sulphur either dusted or sprayed is a specific against it.

On the West Coast several other species of mite occasionally cause considerable damage to foliage. For some, sulphur dusting is the specific; for others, one or another of the contact insecticides must be used.

Nematodes. There are many species of these microscopic eel-like creatures which feed on vine roots. They cause comparatively little damage to the native American varieties and French hybrids grown in the North and East. But in the hot and humid parts of the South, in parts of Mississippi and Texas, they can make the growing of certain varieties practically impossible. They can also be terribly destructive to *vinifera* grown in the hot, sandy soils of California. Grafted *vinifera* vineyards encounter much less nematode trouble, and certain rootstocks are definitely nematode resistant. See page 195.

Birds. As the people of India consider the cow a privileged creature, so most Americans are brought up with a protective feeling toward birds. We are taught that birdsong is beautiful[1] and most of us think we should enjoy it even though we don't, even though it interferes with our sleep in the cool of the predawn hours. The thought of *killing* a thrush or robin redbreast in cold blood is repugnant, and we tend to consider the fondness of cats for bird-hunting a serious defect in their character. When we travel in Europe we are therefore puzzled and shocked at the enthusiasm of

[1] See Rachel Carson's *Silent Spring,* an evocation of a world without birdsong.

Frenchmen and Italians for shooting birds of all shapes and sizes, including the smallest.

The explanation is that birds eat grapes and if not themselves shot and eaten will reduce the vintage considerably, because birds enjoy grapes at a stage of ripeness just before they are fit to be picked for wine. The wine-grower who has struggled all season to bring through his crop does not surrender it willingly to birds, no matter how beautifully they sing, and he knows there will be lots more of them next year no matter how many he shoots on the sly. Birds and wine-growers are natural enemies, like dogs and cats.

This has been a matter of record since the time of Cato, who was definitely an anti-bird man and when not saving Rome spent much of his time devising ingenious but unsuccessful ways of trapping and killing them. And it is the small grower especially, whose vineyard may have trees nearby or even be up against forest or woodlot, who suffers the greatest damage. What is new about the situation is that birds, like human beings, have lately been undergoing a population explosion. They are on the increase. No longer is it only the small isolated vineyard that suffers their depredations. They are now a Grade A nuisance in solidly planted viticultural areas and they take a tremendous toll of marketable crop. The starling may be the most destructive species, but others are almost as bad.

Poisoning is ruled out by the moral indignation and legislative activity of the formidable ladies of the birdsong lobby, and so is the sport of thrush and ortolan hunting as enjoyed in France and Italy. Most efforts at control do little more than shoo birds from one appetizing vineyard to the next. This is the effect of firecrackers, acetylene cannon, phonograph records of the starlings' warning cry, stuffed owls hung in conspicuous places and intended to frighten, helicopters, balloons with mechanical noisemakers, rockets, plastic and metal whirligigs strung on cords, blank cartridges, arm-waving, popguns, gongs, dishpans, scarecrows, and an infinity of other feeble defenses.

The small grower can cover his trellis rows with strips

of tobacco cloth and save his crops. In Switzerland they make a plastic substance that comes in long skeins of filaments finer than those of a high-priced wig. This material may be spread out into a sort of cobweb of almost indefinite width to cover trellis rows. The birds entangle their wings in it, hate the sensation, and depart as soon as they can disentangle themselves, leaving the crop almost intact. But such protections are expensive and are impractical for commercial vineyards.

Soon or late, grape-growers are going to have to unite with cherry-growers, strawberry-growers, rice-growers, and other such victims of pro-bird bias (it is they who are silent, not the birds) and conduct sit-ins and other demonstrations demanding changes in public policy.

Other Pests. Though the principal enemies have been mentioned, the list has by no means been exhausted. Cutworms can damage the shoots of young, newly set vines. Rabbits are fond of young vines too, and deer love to browse on grape foliage and easily jump fences to get at it. White ants (termites) sometimes move into the trunks of old living vines in California's Central Valley and hollow them out. The sphinx or hawk moth has larvae which resemble the familiar tomato caterpillar and like grape leaves. Then there are the grape curculio and the mealy bug; there are root borers of numerous species. One might mention aphids, click beetles, the lead-cable borer, scale insects, and the false chinch bug. But why go on? When something strange shows up, there are always those public servants the county agent and the state entomologist. Let them scratch their heads and read up on it.

3. OTHER AILMENTS

There remains the miscellaneous category of "other ailments." Some have a climatic origin, some are caused by virus or bacteria, and some are nutritional.

Climatic Ailments. These have been mentioned earlier. Spring frost injuries are prevented by choosing a non-frosty site—a site providing good air drainage; by power propellers and other devices for stirring up air on frosty nights; by covering the vines with mats or cheesecloth when frost threatens (expensive, and not practical for commercial vineyards); by heating devices; by training the vines high when they must be planted in frosty locations; and by choosing varieties capable of putting out a second crop after frost has ruined the first one.

As for winter killing, there is nothing to do but to cut the vine back one or two nodes below the killed part. If the vine is killed to the ground, the dead trunk should be sawed off and the vine retrained from a sucker of the next season. A precaution used in some cold winter areas is to train vines to two trunks, on the chance that one will survive even though the other may be killed back. It is normal for some of the vine's cane growth to be killed back each winter. That doesn't matter, since most cane growth is destined to be pruned off anyway.

Sunburn causes little damage in the East. In California it is apparently promoted by drought conditions as well as by direct exposure of the fruit to the hot sun. The best means of avoiding it are irrigation, the use of drought-resistant stocks, training methods that shade the fruit, and summer pruning to encourage the growth of lateral shoots.

Bacterial and Virus Ailments. Crown gall, called black knot in California, is seen when alternations of mild and freezing weather sometimes cause cracks in trunk or arms of a vine. The organism *A. tumefaciens*, always in the soil, enters the lesion and develops cancer-like swellings or galls at points of injury. If the infection is light it eventually withers and disappears. If there is much of it, the thing to do is cut the trunk below the point of injury and train up a shoot of the season to replace it.

Several different virus diseases have caused serious trouble in California. The symptoms are malformations of foliage and shoots, a sharp drop in vigor and productivity, and eventually the death of the vine. In the case of *fan-leaf virus* the malformation of leaves is practically identical with that caused by exposure to the dangerous weed-killer 2, 4-D.

Before assuming that virus is present, make sure that a near neighbor hasn't been using 2, 4-D on a cornfield or hedge-row. A little of it, drifting in a breeze, provokes a lot of symptoms.

The only thing to do on evidence of virus infection is to call in the specialists; and if the symptoms are confirmed they will tell you to uproot the infected vines as soon as possible. The way to avoid virus is to avoid virus-infected land—i.e., use land not previously populated by diseased vines—and to use only wood from healthy vines in propagating. State-certified virus-free material is now available in California. Virus disease has not up to now been a problem in the East.

Nutritional Diseases. Three of these, to which some reference has already been made, are chlorosis, coulure, and millerandage. The symptom of chlorosis is premature yellowing of the leaves, brought on by a lack of chlorophyl, and it is usually associated with too high a concentration of lime in the soil. The remedy is to use lime-tolerant root stocks. Potassium deficiency and deficiency of some trace elements are also indicated by discoloration and withering of foliage. Salt damage, caused by high concentrations of chlorides, and alkali damage, caused by poor irrigation practices, are characteristic of desert regions.

Coulure and millerandage are phenomena resulting from defective fertilization of the blossoms. Millerandage is the setting of defective little berries, and coulure is the failure to set at all. Some varieties show these defects regularly because of defective flowers, and the cure is to interplant a few varieties potent in pollen. But any variety may be subject to coulure when the vine is too vigorous (as when it has been overfed with nitrogen) or when weather conditions at blossoming time are unfavorable—that is, too cold or rainy.

4. A ROUTINE FOR DISEASE CONTROL

The catalogue of major and minor vine diseases is formidable. Winkler lists 50, and his count is probably conservative. But it must be remembered that they are not all present everywhere and that they ebb and flow. Black rot, downy mildew, or leaf hoppers may be serious one year and almost entirely absent the next. So the grape-grower isn't fighting all of them all of the time. Remember also that in the discussion certain remedies were repeated over and over. In medicine the general practitioner prescribes the same remedy for more than one ailment and whistles for a specialist when he encounters something beyond his competence. So the grape-grower has certain sovereign preventives and remedies that he applies in regular routine: the copper sprays, sulphur, certain of the new organic fungicides and insecticides. He holds to a regular schedule—that is vitally important—and if that doesn't keep the vineyard healthy he goes to his books or calls for help.

In California the basis of the routine is dusting with sulphur against the prevalent powdery mildew. Winkler in his authoritative treatise[1] recommends six dustings at the rate of 5 to 10 pounds per acre applied at temperatures below 95°, timed as follows, and to be repeated in case of heavy rains:

1. When shoots are 4 to 8 inches long
2. When shoots are 12 to 15 inches long
3. Fourteen days after second dusting (about blossoming time)
4. Fourteen days after third dusting
5. When berries are about half grown
6. Just before fruit begins to ripen

For the control of other diseases he does not prescribe a fixed routine, since the University annually revises its recommendations on control materials and procedures. Growers in California may obtain the annual revision of *Leaflet 79* from their farm adviser or by writing the Public Service Office, College of Agriculture, Davis, California.

In the East the traditional basis of disease control is Bordeaux mixture, which is effective against the three im-

[1] A. J. Winkler: *General Viticulture* (University of California Press; 1962).

portant fungus diseases, downy mildew, powdery mildew, and black rot. It consists of either copper sulphate or preferably another copper compound known as *fixed copper*, plus slaked lime, plus water. These are usually mixed at the rate of 3 pounds of the copper compound and 6 pounds of lime to 100 gallons of water, though some prefer the stronger 4-8-100 proportions. Proportions need not be precise, but the quantity of lime should always be twice that of the copper compound.

For a small domestic vineyard one may use "prepared" Bordeaux, which consists of ready-mixed copper sulphate and lime, and mix this with water as required, but it should always be prepared fresh.

It is just as easy, and cheaper, to begin with the separate ingredients, as large growers do—a bag of fixed copper or copper sulphate powder and a bag of best-grade builder's lime. The method of preparation is the same whether a big tractor-powered sprayer or a 4-gallon garden sprayer is involved. First fill the spray tank three-fourths full of water. Then dissolve the correct quantity of copper (whether pounds or spoonsful) in a plastic bucket and add to the water, stirring. Then make a paste of the correct quantity of lime in the bucket, adding water slowly and stirring. When the paste is smooth, dilute to the consistency of milk, still stirring, pour this in after the dissolved copper compound, agitate thoroughly, and apply to the vines. The lime should be screened into the tank to keep out lumps likely to clog the spray nozzle, and the mixture should be kept agitated while applying to keep the lime in suspension.

Remember that Bordeaux mixture, or copper spray, is a fungicide—*it fights the fungus diseases only and does not destroy insect pests*. However, it does serve as a *carrier* for the insecticides where and when they are necessary. With the proper insecticide added in the proper quantity, the Bordeaux mixture becomes both a fungicide and an insecticide and thus saves the grape-grower extra trips and labor.

In the East the favored insecticide is DDT because it

SPRAY CHART

(Based on 100 gallons. Reduce quantities proportionately for less.)

When	*Why*	*What*	*Notes*
1 Shoots 1 inch long; repeat when 6 inches	Dead arm Flea beetle	2 lb. Captan or Folpet, 50% WP 2 lb. DDT, 50% WP	Apply only if dead arm is a problem. Use DDT only if flea beetles are present.
2 Early pre-bloom, about 10 days before pre-bloom	Black rot Downy mildew	3 lb. fixed copper *plus* 6 lb. lime *or* 2 lb. Captan	Needed only in problem vineyards.
3* Pre-bloom, as blossoms start opening	Black rot Downy mildew	Same copper-lime, *or* 2 lb. Captan or 2 lb. Folpet	Indispensable in most Eastern vineyards.
4* First post-bloom, when most blossom caps have fallen	Fungus diseases Leaf hopper Berry moth	Same copper-lime, *or* 2 lb. Folpet 50% WP 2 lb. DDT 50% WP Spreader-sticker	Indispensable in all Eastern vineyards.
5* Second post-bloom, 8 to 12 days later	Fungus diseases Leaf hopper Berry moth Japanese beetle	Same copper-lime, *or* 2 lb. Folpet 50% WP 4 lb. micronized sulphur 2 lb. DDT Spreader-sticker	Indispensable in all Eastern vineyards. Use sulphur only for vinifera or hybrids badly affected by powdery mildew. Use sulphur only with copper-lime, not with Folpet.
6 Repeat sprays (at 12 to 14 day intervals as necessary, until ripening begins)	Fungus diseases Berry moth Japanese beetle Leaf roller	Same copper-lime *plus* 4 lb. sulphur, *or* 2 lb. Folpet 50% WP 2 lb. DDT or methoxychlor *or* Sevin 50% WP Spreader-sticker	Substitute Sevin for DDT in final sprays because it disintegrates quickly. No spreader-sticker in late sprays.

provides routine protection against most of the prevalent insect pests. The preferred form is the so-called 50 per cent Wettable Powder (WP), used at the rate of 2 pounds per 100 gallons. The proper quantity is mixed to form a smooth paste, which is then thinned and poured into the spray tank after the copper and the lime. Such minimal and selective use of DDT and the other new organic pesticides is not to be confused with the wholesale use of them for blanket or area spraying which has aroused so much concern in recent years.

This then is the basic vineyard spray formula. Except in the most favored circumstances, three applications are absolutely essential. These are the *pre-bloom*, applied just as the grape flowers begin to open; the *post-bloom*, applied as the blossoming period ends and most blossom caps have fallen; and the *second post-bloom*, applied eight to twelve days later. These are indicated on the chart with an asterisk (*). One or more earlier sprays may be useful in special circumstances, as the chart indicates. Subsequent applications also depend on special circumstances. Thus a rainy summer will require subsequent applications at twelve-day or two-week intervals up to the beginning of ripening; and in areas subject to the Japanese beetle and some other mid-summer destroyers the use of an insecticide will have to be continued, but with the substitution from late July onward of an insecticide less persistent than DDT.

See the chart for detailed instructions.

5. SPECIAL SPRAYS

In addition to the combination of DDT for insect control and copper spray for fungus-disease control, there exist a number of new insecticides and fungicides. These are complex organic compounds of great potency, and their number is constantly being increased. They provide means of tailoring the spray program to special needs, and well-informed commercial growers make a great deal of use of them. Amateurs may not care to bother with them unless they undertake the adventure of growing the *vinifera* varieties, which are so much more susceptible to disease than the native American varieties and most of the French hybrids.

Of the fungicides, *Captan* is effective against black rot and downy mildew, but not effective against powdery mildew. There is no time limitation on its use.

Ferbam is effective against black rot only. It provides no protection against downy mildew or powdery mildew. A time limitation of seven days between final application and harvest should be observed, if it is used.

Folpet (Phaltan) is effective against all three fungus diseases, black rot, downy mildew, and powdery mildew. It can therefore replace copper spray throughout the program. It is more expensive. On the other hand it does not clog spray nozzles as copper-lime spray does, and it does not leave a conspicuous spray residue. No time limitation.

Zineb (Dithane) controls downy mildew and is used by a few New York growers in combination with Ferbam as an early spray.

Sulphur, the staple in California to control powdery mildew, is also useful in the East, but as a spray rather than a dust. It gives additional protection against powdery mildew for varieties especially susceptible to it such as the *vinifera* and some of the French hybrids. It may be used in combination with the organic fungicides mentioned above or copper spray, at the rate of 4 pounds per 100 gallons. It burns the foliage of most native varieties and a few French hybrids such as Foch.

Compatibility. The organic fungicides should not be used in combination with copper spray, as chemical changes take place which destroy their effectiveness.

As to insecticides, the old standard, the highly poisonous lead arsenate, is no longer used; nor are nicotine and rotenone usually, though the latter may stage a comeback one of these days. The new synthetics have the field practically to themselves, and of them all DDT is the most generally useful. It has not been shown to build up a dangerous permanent residue if it is prudently and selectively used. However, its persistence causes a residue problem if it is used late in the season. In sprays applied after late July a less persistent insecticide should be substituted. Moreover, some insects can develop an immunity to it with frequent exposure, so that it is sound in principle to vary insecticides once in awhile.

Methoxychlor is effective and may be substituted for DDT, preferably in the earlier sprays. It has a time limitation of fourteen days.

Sevin is as effective as DDT on application. But it loses its effectiveness rapidly, so that it does not give sustained protection. For that reason it has only limited value in the earlier sprays unless its application is timed exactly. Its place is as a substitute for DDT in end-of-season sprays.

Parathion and Guthion, of the new phosphorus group of insecticides, are extremely effective. They are also extremely poisonous to humans and dangerous to use by inexperienced operators. Amateurs should steer clear of them unless they are prepared to follow instructions meticulously. They provide instant kill for insects and they disintegrate rapidly. They have no time limitation.

Spreader-Stickers. One or another of these, such as Dupont or Ortho, added to a spray mixture at the rate of 2 to 4 ounces per 100 gallons, improves both spreading and sticking and extends the effective life of a spray application. One of them should be used in spray applications, as indicated on the chart. It should be left out of late sprays so that spray residue will have a chance to weather off before the crop is picked.

Chapter XI

PROPAGATING THE VINE

Most wine-growers live and die without ever having propagated a grapevine. They leave the risks and delays of this specialized work to the nurseryman, and when they need grapevines they buy them ready-rooted. Grapevines are not expensive, but for the experimental-minded there is a certain fascination in the business of propagation, whether it be done in order to multiply existing varieties or in order to try one's hand at the amusing but not very rewarding work of breeding new varieties.

Seedlings. The seed is Nature's mechanism for reproducing the vine, but so far as man is concerned the growing of vines from seeds is never resorted to in the ordinary course of events. This is because vines do not breed true. Grape seeds cannot be counted upon to reproduce the parent plant, even when the parent plant is one of the self-fertile varieties. The seeds from a domesticated vine that ordinarily bears blue fruit may produce vines bearing blue, red, or white fruit, or may produce male vines incapable of bearing fruit at all, or may produce vines differing from the parent in other equally striking ways. Seeds are therefore planted only by those interested in developing new hybrid varieties, or by those studying hybrids of unknown parentage in the hope that selfed seedlings will provide clues as to the parents. In a word, the growing of seedlings is practically confined today to experts who have made a thorough study of the genetic behavior of the grape and who know what they are after. Amateurs who are tempted nevertheless to try their hand at the rearing of seedlings are referred to the bibliography at the end of this book.[1] But a few words of warning

[1] Munson gives very explicit instructions on hybridization.

ought to be offered: most seedling vines are inferior to their parents in vigor, hardiness or resistance of vine, or in fruiting habit or quality of fruit. Maurice Baco, the French hybridizer, made 50,000 crosses in order to get eight that he thought worthy of dissemination.

Propagation by Cuttings. The usual method of propagating the vine is to plant cuttings of the one-year-old wood, for in this way all the characters of the parent are reproduced in the young vine. Indeed, the young vine is hardly to be considered a new vine at all, but a mere spatial and temporal extension of the parent, comparable to the multiplication of cells by fission.

Wood that is to be used for cuttings should be mature and taken from healthy vines. Neither the fattest nor the thinnest canes should be selected for the purpose, but good straight canes of average thickness. Such canes are cut into pieces from six inches to a foot long, each piece containing two to four eyes. The bottom cut is made just below an eye, and the top cut an inch or so above an eye: the cuts should be slanting. Some nurserymen prefer, if possible, to leave a "heel" of the old wood at the base of a cutting; but except for varieties that root with difficulty this does not seem to offer any real advantage. Cuttings to be used in dry climates ought to be somewhat longer than cuttings for the damp climates of eastern United States, for in dry climates it is desirable to have deep-rooted vines.

It doesn't much matter when the cuttings are taken from the vine, if this is done during the dormant season and before the spring growth begins. Once the cuttings are made, they should be tied into bundles of fifty or so, stratified in moist sand, and kept in a cool place until planting time. Sand, not loam, should be used; and it should be kept moist but not sopping.

The nursery should have fertile, well-drained, well-worked soil, sufficiently deep to allow the easy development of the young roots. A manure dressing, before plowing or spading, is desirable.

Planting takes place in the spring as soon as the soil is

FIG. 22. *Planting cuttings.*

in good condition. A V-shaped trench is dug, or turned with a plow, deep enough so that cuttings may be placed vertically against the side of the trench with the top eye protruding. Cuttings should be placed about three inches apart in the trench; and rows should be far enough apart to allow for cultivation between. When the cuttings have been placed in the trench, dirt is thrown back into the row and well tamped down around them in such fashion that only the top eye is exposed.

Presently the top eye puts forth its young shoot: the new vine has started, but its survival is not yet assured, for the development of the roots, from the nodes below the surface of the ground, usually lags behind that of the shoot. The great danger is that the shoots will put forth so vigorously that the food reserves in the cutting will be exhausted before the new roots begin to function.

During the growing season, the young vines should be kept well cultivated, and should be watered if rainfall is not sufficient. They will make from a few inches to three or four feet of growth, depending upon the variety and cultural conditions. In cold climates, they are best hilled up a little in the autumn. In the following spring they may be dug and transplanted into the vineyard, or if growth has been slow they may be given another year in the nursery before transplanting.

Single-eye cuttings are sometimes used for the purpose of multiplying the number of vines of a given variety as quickly as possible from a limited amount of wood. The cuttings should be made with slanting cuts, so as to leave about an inch of wood above and below the eye, and should be planted about an inch deep in the open or in a hot- or cold-frame.

Propagation by Layering. Varieties differ greatly in the ease with which cuttings strike root – some of them rooting almost 100% and others rooting only with the greatest difficulty. Layering is a method of propagation frequently resorted to for varieties that root with difficulty. It is a scheme for causing roots to grow on a cane before that cane has been severed from the mother vine. The method is simple. A cane of the mother vine, preferably one starting near its base, is bent down and buried so that all but the two end eyes are covered with earth. Shoots develop from the two exposed eyes, and are nourished by the mother vine until new roots develop at the nodes which are buried. After one or two growing seasons, these roots are well developed, and it is then possible to separate the young vine by severing its connection with the mother plant and transplanting it. A variation of the practice is to bend down a mature cane, laying it in a shallow trench containing rich soil and covering the cane full length without exposing the tip. Roots and a shoot develop from each node. When the row of young vines has attained sufficient growth, they are cut apart and transplanted. If a vine is to be used for layering, it is usually pruned back severely the previous spring in order that especially strong canes may develop close to the ground.

Propagation by Grafting. Grafting is an ancient method of propagation, used in many branches of horticulture; but in no other has it been more thoroughly investigated than in grape-growing. Essentially, grafting is the operation of joining the parts of two plants in such fashion that the roots of one will support and nourish the other. Or, to put it in another way, it is a method of providing a vine with artificial underpinnings when for one reason or another its own underpinnings won't do.

In grafting, the plant that provides the root is known as the rootstock, or simply stock; and the plant that it supports is known as the scion.

Grafting is undertaken for a number of reasons. First and most important is the necessity for providing roots for *vinifera* grapes – roots immune to the ravages of phylloxera. It is this urgent need that has been responsible for most of what we know of the theory and practice of grafting.

But grafting has other uses.

It is a good way to propagate varieties whose cuttings do not take root easily by themselves.

Grafting is sometimes resorted to, also, to make over a vineyard from a variety that is unprofitable or otherwise unsatisfactory to one that is more desirable. The new variety thus comes into bearing several seasons sooner than it would if an entirely new vineyard were set out.

Grafting is sometimes used to adapt a variety to soil to which it is not normally adapted. For example, some of the American hybrids that do well in New York State will not grow well in Florida unless they are grafted on special stocks; and they will not grow successfully in the limy soil of Texas unless they are grafted on lime-resistant stock.

Grafting on nematode-resistant and rot-resistant stocks may be used in regions where such root troubles are prevalent. Grafting is sometimes used also for the speedy propagation of a new or scarce variety.

And grafting may also be used in some circumstances to increase the vigor and fruitfulness of certain varieties beyond their ordinary capacity. Thus, it has been found that the vigor of the Delaware may be considerably increased, and its fruitfulness increased also, by grafting it on a vigorous stock.

Bench-grafting. Of the two principal methods of grafting, bench-grafting is the more commonly used by nurseries. The necessary materials are (1) a knife that is really sharp; (2) some raffia; (3) some sound and well-matured dormant canes of the variety to be used as rootstock; (4) some equally sound and well-matured canes of the variety to be used as scion; and (5) callusing box. It is immaterial whether these canes are cut fresh from the vines at grafting time (usually about a month before planting time) or were pruned from

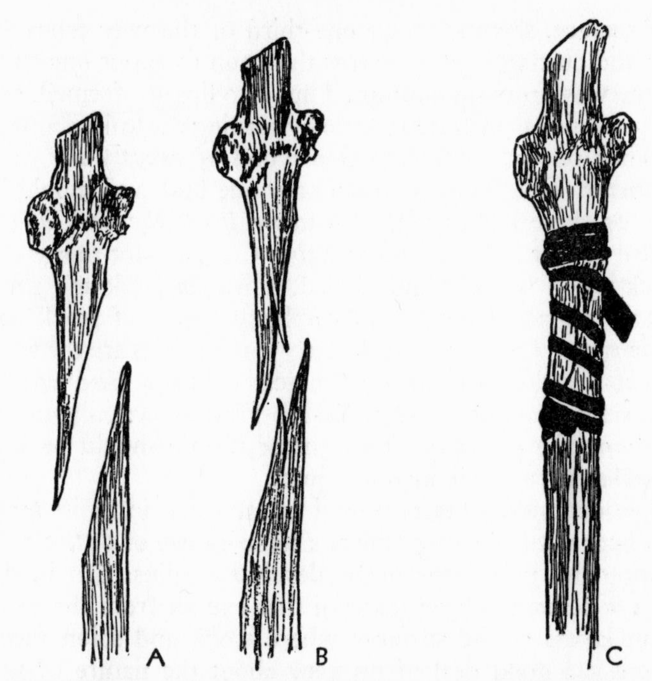

FIG. 23. *Bench grafting.* (A) *First, make slanting cuts in scion and stock;* (B) *form the slots, or clefts, which are to fit together;* (C) *fit them snugly together and bind.*

the vines the previous autumn and stored in moist sand during the winter. If they have been stored, they should be rinsed of their grit and soaked for a few hours.

The wood to be used for stock should be cut into pieces 10 to 14 inches long.

The strongest and simplest form of graft is that known as the English graft or whip graft. This is the one that will be described.

First take a cutting of the wood to be used as stock and cut off the eyes. Then, an inch above its top node, make a clean sharp cut at an angle of 45° (some prefer a longer cut than this as shown on this page). This cut must be perfectly smooth and true, and is best made with one continuous stroke of the knife. Then make a slanting cleft in this freshly cut surface, starting about one-third of the way from the top and cutting slightly across the grain to about one-third of the way from the bottom. Thus two lips are formed, and they should be separated a little by turning the knife slightly.

Then select a piece of scion wood *of exactly the same diameter*. This should contain only one bud, and should be two or three inches long. *Do not rub off the bud, as the entire vine is to develop from this.* Cut and slot this as the stock wood was cut and slotted. Then fit stock and scion snugly together so that the cambium layers of stock and scion meet. They will fit perfectly if the angles are the same and if the slots are made as indicated. Then bind with a few snug turns of moist raffia. No grafting wax should be used: on the contrary, the turns of the tie should be separated so as to admit air to the joint.

The completed grafts should be stratified in moist sand, and kept until planting time at a temperature of 65° to 75°. During this period the grafts develop a callus: that is, the joint is cemented by a material that exudes from the cambium layers at the surfaces where stock and scion meet. There is a good deal of mystery about the nature of this callusing process; but it need not bother us here—the important point is that a healing tissue is thus developed. The callus develops at variable speed, depending mainly on the temperature at which the grafts are kept. Below 60° it does not develop. At 65° it is well developed in a month. In California it is usual to employ a mixture of sawdust and coarse charcoal for the callusing instead of sand. Or sawdust alone or sphagum moss will do.

By planting time, the grafts should be well callused. The scion eye will usually be found to have started, but this does not matter so long as the swollen bud or tender little shoot is not injured. Tiny roots will also have started from the stock, and usually some little roots from the scion near the point of union; scion roots should be rubbed off, but not the others.

The nursery should be prepared as for the planting of ordinary cuttings. The grafts are placed vertically against the side of a trench, several inches apart, and well heeled in. But the scion eye is not left exposed: instead, the whole row is hilled over with earth in order to prevent the tender joint from drying out and to protect it from sudden changes of temperature. The earth should cover the tops of the scions about two inches.

The hilled earth should not be disturbed until the young shoot from the scion eye has pushed through and made considerable growth. Then the row may be lightly cultivated. During the summer, the row should be carefully cultivated, and watered if necessary. Toward the end of July, the hills of earth should be lowered and the graft joints examined. Scion roots will have formed at the joint, and these must be rubbed off or cut off. Earth is then hilled around the joints again.

In the autumn, when growth has stopped, the young vines should be well hilled for the winter. They are ready for transplanting in the vineyard the following spring.

Bench-grafting is also undertaken on one-year-old rooted cuttings instead of unrooted cuttings. The making of the actual grafts is the same, but as soon as they are grafted they may be returned to the nursery. Formation of the union takes place under the mound of earth. Before they are returned to the nursery, the roots are cut back to two or three inches; and the rooted cuttings are planted somewhat farther apart than the unrooted cuttings, since root-growth will be much greater.

Field-grafting. This method of grafting is undertaken on stocks that are rooted in the ground, in vineyard or nursery.[1]

When the stocks to be grafted are more than a year or two old — when, for example, a bearing vineyard is to be made over from one variety to another — whip-grafting (see page 167) becomes difficult, for the reason that the stocks and the scion wood to be grafted upon them are not of the same diameter. Hence the procedure used is quite

[1] A special form of field-grafting, known as bud-grafting, is used a good deal in California. The process is well described in farm circulars. There also exist machines for both bud- and bench-grafting.

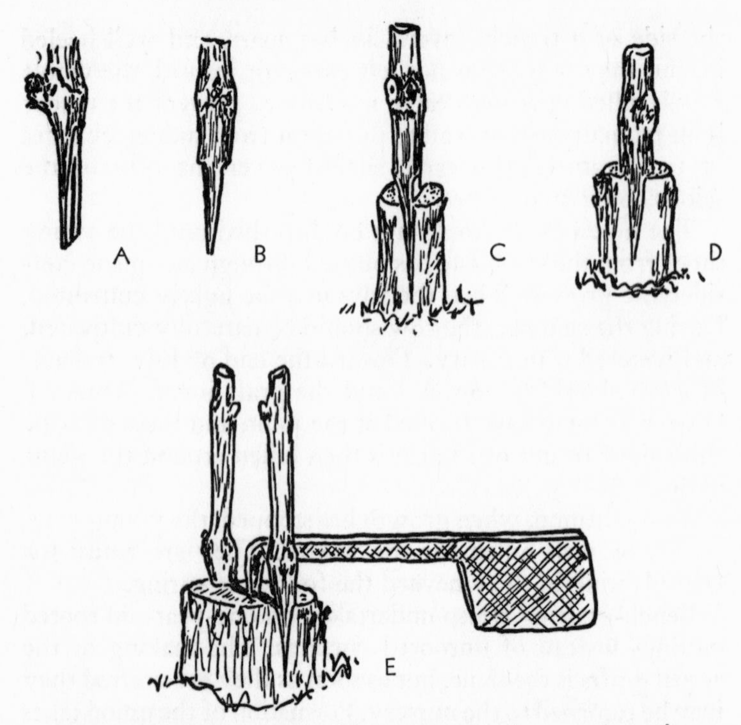

Fig. 24. *Field-grafting.* (A) *conventional way of cutting scion wedge;* (B) *scion wedge cut with shoulders;* (C) *and* (D) *fitting scions into stocks;* (E) *fitting two pieces of scion to the crown of a mature, vigorous stock, thus providing the stock with four buds into which it may pour its energy and reducing the production of unwanted suckers.*

different; the method here described is exactly the one used and recommended by Cato in the second century B.C. It is usually called the cleft graft.

The vine to be grafted should be cut off just below its crown where the grain of the wood is straight, at or just below the surface of the ground. This should be done several days before the actual grafting, in order that the heaviest flow of sap from the wound will be over when it is time to graft.

Make a one- or two-bud cutting of scion wood. Starting just below the bottom eye, whittle it to a clean, narrow wedge about an inch long. Then, with a sharp, thin chisel or grafting tool, split a cleft in the top of the stock, being careful not to split too deeply. Keep the cleft slightly sprung with a corner of the chisel, and insert the wedge-shaped scion so that the cambium layer on one side of the wedge coincides with the cambium layer of the stock. When it is snugly in place, remove the point of the chisel from the cleft and it will close on the scion, holding it firmly in place. No tying is necessary if the stock is more than an inch in diameter. In cutting the scion to fit into the cleft, it is best to leave one side of the wedge (that side which will coincide with the cambium layer of the stock) a little broader than the other, so that the cleft in the stock may close as completely as possible.

When the scion cutting has been firmly seated in its cleft, the entire graft is mounded with soil. No grafting wax is needed. The mound of soil should cover the scion well, and it should not be touched until a shoot from one of the eyes of the scion is well advanced; for the new union is delicate and easily broken.

The chief difficulty with grafts on large stocks is that the stocks do not find a sufficient use for their vigor in the two buds of the scion. In consequence, they send up shoots in enormous quantities from adventitious buds located below the graft on the stock. These grow with great vigor — almost, it seems at times, fast enough to be seen in action — and the danger is that the stocks will pour all their energy into these suckers and thus leave the scion to languish and die for want of sustenance. Yet the suckers cannot be hastily removed for fear of breaking the tender new union between stock and scion. All that can be done is to cast a balance between possible injury to the new scion and the danger of allowing the suckers to go too far. The best compromise is to wait until shoots from the scion are making good growth, and then, on the first rainy day, or at any rate when the soil is moist and soft, to remove a little soil from around the graft, make sure that the suckers are not wound around the scion, and yank them off.

During the first season, the mound should be maintained around the union in order to keep it from drying out. In subsequent seasons, it may be dispensed with.

Green-grafting. Of the many other forms of grafting, only one more will be described: green-grafting, a method that is of interest only to amateurs. It is not adapted to mass-production nursery practice, but it is very certain in its results.

Green-grafting is undertaken in middle or late June when the shoots have attained much of their growth and have begun to stiffen (but are not yet woody), and when the bud-eyes of the next season have begun to form on them. The procedure is simple, though difficult to describe. Cut a short length of green cane, containing one node, from the vine that is to be used as scion, and remove the leaf. Then select a good healthy shoot on the vine that is to be the stock. Remove all leaves from it, and cut it with a slanting cut about an inch beyond the fourth node, counting nodes from the base. Bring together the slanting cut of the cane to be used as stock and that of the short length of green scion. Then wrap this simple lap joint with a cigarette paper, and tie the cigarette paper in place with a piece of rubber band. If the cuts are cleanly made and snugly fitted, sap from the mother vine will promptly find its way through the joint and begin to bleed from the end of the scion. Before long, a perfect union will have been formed, and the cane will continue its growth (from the scion node) as though nothing had happened – *except* that we now have a cane one-half of which is stock and the other half of which is scion. The cigarette paper is merely to protect the union from drying out until it has healed (healing proceeds rapidly and surely with green-grafting), and the rubber band holds the joint tightly yet allows the growing shoot to expand. When the dormant season arrives, simply prune this cane so as to obtain a cutting that is half stock and half scion, and plant this cutting in the spring. When only a few grafts are wanted, this is a simple and easy way to make them – and the union is as nearly perfect as one can be.

Theory of Grafting. So much for the actual methods. Now we may look into some of the peculiarities of what is, after all, an unnatural association. When grafting came under discussion as the most promising method of reconstructing Europe's devastated vineyards, certain men denounced the practice with great heat and show of learning. They contended that it would only substitute one kind of disaster for another. Stock and scion, they contended, influence one another reciprocally; and these reciprocal influences would cause permanent modifications in the traditional grape varieties – such disastrous modifications, for example, as the acquisition of the "wild" or foxy flavor of the American grapes. Yet, as a matter of fact, none of the dire predictions of the conservatives has come true. The view today is that certain reciprocal influences between stock and scion do indeed exist, but that these influences give rise to no fundamental alteration in the nature of either. Rather, the influences are of the same order as the influences of soil, cultivation, and climate. They affect *growth* only – not the inherent character of either stock or scion. The prevailing view today is that even these influences may be brought to a negligible point by grafting scions only on those stocks which are known to be congenial to them.

Congeniality is not here used in a vague or mystical sense but to express the concrete behavior of a grafted vine. A successful grafted vine never becomes one individual, but remains an association of two different plants functioning more or less successfully. Congeniality – or affinity – is the word used to express the degree of success of that association; and perfect congeniality occurs, as Husmann says, only when "a variety grafted on another behaves as if the stock were grafted with a scion of itself, the union being perfect and the behavior of the vine the same as that of an entire ungrafted plant." It became apparent very early in the investigation of grafting that all species and varieties otherwise suitable as rootstocks are not suitable for all scions. The work of fitting the most congenial stock for any given scion variety is slow, however, and has not yet reached any great degree of refinement.[1]

Since perfect affinity is never achieved, stock and scion usually do show the results of reciprocal influences. Thus a scion grafted on an exceedingly vigorous stock (such as the Rupestris St. George, which is the "standard" stock of California, or the Mourvèdre x *rupestris* 1202, which is widely used in France) will grow with abnormal vigor. Conversely, a normally vigorous scion grafted on a weak-growing stock will find its vegetative vigor much reduced. In both cases, the effect is much as though the scion had been growing on rich and humid, and on thin and dry, soils respectively. The other effects of grafting that have been noted from time to time – the increase or decrease of crop, the advancing or retarding of growth in the spring and of ripening dates, the increase or decrease of the crop's sugar content, the increase or decrease of hardiness and resistance to disease – may all be explained in the same way.

Likewise, it is now generally admitted that the only differences between the fruit – and therefore the wine – of grafted and of ungrafted vines is that which is directly traceable to congeniality and vigor; that is, differences in the health of the vine and in the perfection with which it ripens its fruit. The nature of the fruit, aside from such variables as sugar and acid content which vary from year to year anyway, is not altered.

[1] Descriptions of the most generally used rootstocks will be found on pp. 195–7.

Chapter XII

GRAPES FOR CALIFORNIA

In this chapter will be found descriptions of 60 *vinifera* grape varieties – 36 red-wine varieties and 24 white-wine varieties. Considering that some 6,000 to 8,000 varieties of *Vitis vinifera* have been described, and considering that most of them could be expected to adapt themselves after a fashion to California conditions, this may seem a drastic feat of condensation. To allay the fear that something important has been left out, let me hasten to say that, of all the varieties that have been described, probably not more than 400 are cultivated on a very large scale anywhere in the world. Of these, perhaps one-quarter account for the greater part of the world's wine. When we come down specifically to the viticulture of California, it turns out that a dozen varieties (and those not necessarily the best for their purpose) provide more than 90% of the State's wine production.

Thus it turns out that a list of 60 varieties is in fact a more varied list than is generally available from the nurseries of the State, and that, when variations in wine quality caused by climate and soil are also considered, it is a basis for a much greater diversification of wine types and characteristics than the State at present offers commercially. It could easily be expanded – and doubtless will be, as time and experience dictate. But for the time being, it represents a fairly satisfactory compromise between a too limited list and one so comprehensive that it would only confuse the prospective grower.

This list is in no sense an original selection. It represents the combined wisdom of those investigators who have labored during all these years to winnow the good from the less-good, the so-so, and the bad, beginning with the work of Hilgard and depending most of all on the contemporary, and continuing, work of Amerine and Winkler,[1] from whom I have borrowed freely. In addition, I have constantly checked the conclusions of these investigators against European experience, the conclusions of European students, and (in the case of many varieties) personal examination in the regions of their origin. In short, the reader may accept these evaluations with confidence.

Three other points: (1) since the list is primarily for practical growers, I have not burdened my descriptions with ampelographic details but have confined them to those characteristics which give each variety its special interest; (2) certain varieties included in the list are *not* recommended, but are included as warnings against their culture, or because they have played an important role historically. The varieties most highly recommended have an asterisk (*) before their names; (3) in each description will be found a regional recommendation expressed in Roman numerals (as Regions I, II, etc.), indicating the region to which that particular variety is best adapted.[2]

RED-WINE VARIETIES

**Aleatico*. An Italian variety with pronounced muscat aroma and flavor, suitable for the production of a lightly colored, natural sweet wine. Its sugar content is too high, and its acid too low, for dry wine; and in any case its pronounced aroma and flavor seem to demand sweetness as an accompaniment. This is a long-season, hot-weather grape, and should be grown only in Regions IV and V. Head training.

Alicante Bouschet. Midseason, heavy producer. This is much grown in southern France and Algeria for cheap blending wines, and it had its great vogue in California

[1] M. A. Amerine and A. J. Winkler, *op. cit.*

[2] The regions referred to are those described in Chapter IV, pp. 50–3.

during the years of Prohibition, when thousands of carloads were shipped east for the domestic wine-makers. Its fruit withstood shipment better than most other varieties — almost its only virtue. The wine has intensely dark color (which, however, is dropped rapidly with aging), a good balance of sugar and acidity, and a slightly unpleasant flavor. The Alicante Bouschet is now a drug on the market in California, and its acreage is likely to drop in the years to come.

Aramon. Another grape to be avoided. It produces tremendous crops of large-berried bunches, the fruit being low in sugar, low in acid, and low in color. It produces a thin wine, pale in color and low in alcohol, the basis for most of the French *vin ordinaire*, usually being blended with grapes of better character. Alone, its wine has little quality, and it has no apparent place in California viticulture.

**Barbera.* Best known of all the Italian red-wine varieties, and very well adapted to Regions III and IV. Barbera has a distinct aroma, and its wine is heavy-bodied, well balanced, and when well aged has a great deal of character. Vine is vigorous and above average in productivity. In Regions I and II it tends to be a bit too acid. In Regions III and IV the fruit attains ideal balance normally. In Region V, despite the handicap of excessive heat, the quality of its wine is above the average. Spur pruning.

Beclan. A variety grown to some extent in northeastern France. It is adapted only to the cooler parts of California (Regions I and II). Elsewhere it qualifies neither as a producer of superior wine nor as a bulk producer. And even in the cooler parts, owing to its limited productivity, better varieties should be given first choice.

Bolgnino. Regions III and IV. Late. Vine of average vigor and above average productivity. This is an Italian variety, yielding a rather rough wine which will improve somewhat with age. In the hot districts it makes better red wine than most of the standard varieties, but is inferior to Barbera.

Buonamico. One of the true Chianti grapes. See San Gioveto.

**Cabernet Sauvignon.* In the great viticultural region around Bordeaux, this grape dominates all the best vineyards. Many of the very finest Château clarets are wine of the Cabernet Sauvignon exclusively. It is to the distinctive aroma of this grape, ripening into a marvelous bouquet, plus the equally distinctive flavor, that the French clarets owe their world-famous reputation.

In the right parts of California, the Cabernet Sauvignon is a highly vigorous vine, and when properly pruned is moderately productive. It is one of those *vinifera* varieties which require cane-and-spur pruning rather than mere spur pruning, though with age and the multiplication of spurs the canes may be eventually dispensed with without sacrificing production. Fruit grows in moderate-sized clusters of small, tough-skinned berries, and is relatively resistant to the vine diseases. The variety is best adapted to Regions I, II, and III, where its wines, when well fermented and well aged (Cabernet always requires aging, to tone down its tannin), have a character strongly resembling the Cabernet wines of the Bordeaux district, though inclined to be heavier. Beyond question, the Cabernet has a great future in California viticulture; and at present it accounts for most of those California wines which genuinely deserve to be called "fine." In the warmer regions, the wine of Cabernet loses much of its distinctive character and becomes merely another heavy, rough, rather ordinary wine.

Canaiolo. One of the Chianti grapes. *See* San Gioveto.

Carignane. This is a heavy-producing rather than a fine-wine variety, its wine being characterized by good body, clean flavor, good color, and little or no distinctive character. It is an ancient variety, a native of Spain and much grown on both the north and south shores of the Mediterranean, for blending with less deeply colored varieties. California already has a large acreage; and it will always have a place in the mass production of wines of ordinary quality. It is a vigorous vine, very erect of habit, bearing its fruit in large, shouldered clusters. It is highly susceptible to mildew, and is therefore suited only to dry regions. It achieves its best quality in Regions I and II, and produces huge crops (but wine of still less character) in Regions III and IV. Spur pruning.

Freisa. Regions III and IV. Early midseason. Moderate vigor, crops of average size when spur-and-cane-pruned. A thrifty-growing variety, which offers no special disease troubles. Freisa is an old and famous grape of the Piedmont district, where the best Italian red wines come from. It has the great virtue, for California, that the fruit is well supplied with acid even when it is grown in the warmer parts of the State. Its wine has an agreeable special aroma, heavy body, and astringency. In Region V it does not color to perfection. For those interested in a wine of marked character, Freisa is worth trying.

**Gamay*. Regions I and II. Early. Vine of average vigor and moderate productivity, somewhat susceptible to mildew. Fruit is borne in plump, compact bunches of moderate size. This is the famous grape of the Beaujolais region, where it produces a delicious wine of special fruity aroma, softness, and good body. A feature of this wine is that it does not require long aging but is best when it is from one to three years old. In California, in certain restricted localities, the grape produces wines that are entirely characteristic. Unfortunately the Gamay situation in California vineyards is in confusion. There exist several varieties, or clones, of Gamay that differ sharply from one another, not only in their cultural characteristics but in the character of their wine. One of these, though nominally Gamay, appears to be a clone of Pinot Noir. Anyone bent on growing and making wine of the Beaujolais type will be wise to check and double-check the source of his vines.

Grenache. Regions I and V. Midseason. Vigorous vine, heavy producer, resistant to disease. The Grenache is widely distributed throughout the hotter parts of France, and is the basis of several different and distinct types of wine. It enters into the blend of grapes that yield the famous, and very heady, red wine of Châteauneuf du Pape; it is the grape from which the famous *rosé*, Tavel, is made; and on the borders of the Mediterranean, where it attains very high sugar content, it is used to make French "port" and that heavy, syrupy reddish brown dessert wine called Banyuls. It is thus a versatile grape. In most parts of California it attains too high a sugar content for good dry wine, and usually lacks sufficient color for red wine. In Region I it is being used somewhat to make in agreeable *rosé* which somewhat resembles Tavel. In the hot Region V it is very satisfactory for the production of sweet dessert wines. Spur pruning.

Grignolino. Regions III and IV. Midseason. Vigorous and productive, the grapes being borne in big clusters of large berries. This is another of the rather special red-wine varieties from the Piedmont district of Italy. Actually, it hardly deserves to be called a red-wine variety, the color of its wine more closely resembling that of strawberry pop. The Italians are very partial to it, especially when it is slightly bubbly. It is a suitable grape for the warm districts, though it ought to be picked before its acidity drops too far. Another grape for growers who either know their market or want to make a specialty wine with abundant character. Spur pruning.

Gros Manzenc. Regions III and IV. Vigorous vine, moderate producer, rather subject to mildew. This variety, grown to some extent in the South of France, has the virtue of providing ample acidity in hot climates. Further, its wine, though hardly to be called "fine," is distinctly above the average in quality and will improve with age. Its wine is far preferable to Carignane, Zinfandel, and some of the other varieties now widely planted in hot districts, though its production is not so generous. Spur pruning.

Malbec. Regions I and II. Vigorous, but found guilty so far in California of uncertain producing habits. Not troublesome as regards disease. Malbec is an ancient French red-wine variety and is perhaps more widely distributed in France than any other variety capable of producing superior wine. It has many synonyms, of which the name Côt is most frequently encountered. By tradition it is blended with the Cabernet Sauvignon, primarily to soften the wine of the latter and reduce the length of time required for aging. And throughout the vast Bordeaux red-wine area, in those less known districts whose wines are marketed simply as *Bordeaux supérieur* and not under Château bottlings, it is probably more widely planted than the Cabernet itself. In the Loire Valley it is the source of such delightful second-best wines as Chinon and Bourgueil.

Merlot. In discussions of red Bordeaux wines, the variety Cabernet Sauvignon usually gets all the credit for their excellence. In fact, Cabernet Sauvignon is but one of a family of closely related varieties differing in fairly important ways but all bearing a family resemblance. Malbec, mentioned above, is one of these. Merlot is another, and in the Médoc there is probably a larger acreage of Merlot than of Cabernet. Its particular virtue is its ability in a blend to soften the austerity of Cabernet wine, bringing it sooner to the point of bottling without sacrifice of quality. The simple truth is that blends of these two in California make wine more nearly resembling red Bordeaux than Cabernet Sauvignon alone. Two wineries have discovered and are acting on this. Does well in Districts 1 and 2.

Mission. Regions IV and V. Vigorous vine, producing heavy crops in midseason. The Mission is the grape, already discussed at some length in Chapter II, on which California viticulture was founded. It is a variety of unknown origin, brought first into California in the 17th century by the mission fathers and by them passed along, as cuttings, from one outpost mission to the next. In the past, it was responsible for an enormous output of very mediocre red wine. Actually its color is not sufficient, and it develops too much sugar and too little acid, for good red wines, these being always flat, heavy, and without distinction. In recent years it has come into its own in an entirely different role. Grown in the hottest districts, its high sugar and low acid become virtues in the making of sherry. Though its sherry is by no means the finest, neither is it the worst; and because of the productivity of this variety, and its long tradition, it is now probably a permanent inhabitant of California's vineyards. Spur pruning.

Mondeuse. Region II. Midseason. A reliable producer, but should not be allowed to overbear. This is another of the better grapes from the hot wine-growing districts of southern France – and, as is usually the case, its desirability lies with its relatively ample acidity. In California, however, the quality of its wine offers no special attractions when it is grown in the hot regions. In Region II it is useful, in blending with ordinary red wines, to give them a bit of zip. But that is about all it seems to be good for in California. Spur pruning.

Nebbiolo. Regions III and IV. Moderate vigor, and requires short cane or long spur pruning for adequate crops. The Nebbiolo is responsible for the very best wines of the Piedmont district of Italy, wines which at their finest can rank with the better Clarets and Burgundies, though they resemble neither, being highly characteristic in bouquet, rather more tannic, and possessed of a special indescribable lusciousness. The famous Italian wines Barolo and Gattinara are made from this grape. In California the grapes tend to lack color, especially in the hotter districts. And on the other hand this vine does not seem well adapted to the very cool districts. In Regions III and IV it deserves to be planted by those not merely content with quantity; and the chances are that it will be rewarding.

Negrara Gattinara. Region III. Midseason, vigorous, amply productive. A vine deserving more trial in this region of relatively high temperatures, by those who require fairly high production yet want to make wine a cut above the ordinary. Quality of its wine is inferior to that of Nebbiolo.

**Petite Sirah.* Regions I, II, and III. Early midseason. Prune to long spurs. Vigorous variety, and a better than average producer. But its smallish, compact bunches rot easily in wet weather; and the fruit is subject to rather wide variation in quality according to the nature of the season. In very hot years the grapes raisin rapidly and the resulting wines are heavy and flat. Petite Sirah has a history, in the Rhone Valley, extending back to Phoenician times, the Phoenicians apparently having brought it into Europe by way of what is now Marseille and the mouths of the Rhone. Wines of the Côtes du Rhone are distinguished for heavy body, deep color, a certain roughness, and with age a bold, highly agreeable, and not too delicate bouquet. The Petite Sirah is largely responsible for this character; and in California, grown in the Regions indicated, its wines run true to type. They are not comparable in quality to those, say, of the Cabernet, being "superior" rather than "fine." But those who like them grow very fond of them indeed. Quality is in inverse ratio to the heat of the growing season. In California it is customary to market the wine of Petite Sirah, or blends that include a substantial proportion of it, as "Burgundy." This is wrong. The wines do not resemble true Burgundy, and deserve a better fate.

**Pinot Noir*. Region I. Early. Cane-and-spur pruning. Vine makes rather feeble growth, and, since the fruit is borne in small bunches, cane pruning is necessary to produce a satisfactory crop. Even then, it cannot be called a very good producer. This factor will make no difference to the amateur bent on making the best wine possible. This grape is is not to be confused with a related but inferior grape called Pinot St. George.

This is the famous plant from which all of the finest Burgundies are made. In France it serves another purpose too, being the principal grape from which Champagne is made. For the latter purpose, it is crushed and pressed quickly, so that the colorless juice will not extract pigment from the skins in the course of fermentation. In California, owing to its earliness, the Pinot Noir has a limited area of adaptation. Grown in the hotter districts, its wines are hardly worth the trouble. But grown in the coolest districts, its wines have a recognizable affinity with the fine Burgundies and are indeed among the best wines produced in California. Their only rivals are the wines of the Cabernet. The possibilities of using this grape as a base for fine sparkling wines have not been exploited in California. In any case, the prospect is not particularly encouraging. In France, the use of this grape for sparkling wine is confined to the Champagne region, which is the coolest grape-growing part of France. The growers of the somewhat warmer Burgundy district always make red wine of it, though we may be certain that they would have entered into competition with the Champagne growers had the Pinot Noir proved suitable for that purpose under their conditions.

**Refosco*. Regions II, III, and IV. Ripens in late midseason. Vine of moderate vigor yielding above-average crops of large, compound bunches. Resistant to mildew, though somewhat subject to sunburning of fruit in the hotter regions. This variety, much grown in the Italian valleys of the Po and the Piave, is better known in California as Crabb's Black Burgundy, though its wine is more characteristically Italian than Burgundian. It has good body, a very pleasant aroma which develops into bouquet with aging, and (much to the Italian taste) rather more tannin than most wines. Best of all, it retains a sufficient acidity under all but the hottest conditions found in Region V. The place of this vine in California is essentially that of a compromise, since its yield is almost up to that of the heavy-producing bulk-wine varieties and the quality of its wine is distinctly superior. Spur pruning.

Rubired and Royalty. Two recent hybrids of Alicante Ganzin x Tinta Cao and Alicante Ganzin x Trousseau, respectively. They are high in color, sugar, acidity, and extract and as their parentage indicates are intended primarily to improve the quality of California ports. They may also have possibilities for red table wine in hot Districts 4 and 5.

Ruby Cabernet. Regions III and IV. A cross between Carignane and Cabernet which carries over the wine qualities of the latter to a remarkable extent but is better adapted to high summer temperatures. Oddly enough, it survives the difficult conditions of Maryland, winter and summer, better than Cabernet.

San Gioveto. One of the three principal grapes grown for the distinctive Italian wine called Chianti. The other two are Buonamico and Canaiolo. The three harmonize well, and each of them contributes something essential to the resulting wine. Buonamico contributes bulk, and by itself, in Italy, yields a pleasant, sprightly wine. Canaiolo contributes good color to the blend. San Gioveto is responsible for that very special aroma and the touch of tannin in the flavor which are the peculiar characteristics of Chianti. Of the three grapes, only San Gioveto has been studied systematically in California. There it seems to mature too late for Regions I and II, and to lose its character in the hotter regions. In Region III, it produces moderate crops and yields wine that is entirely typical. That is to say, the wine has the distinctive aroma for which the grape is known and is low in color. It is not, so far as I know, used in any of the wines currently marketed as California Chianti, wines whose sole resemblance to real Chianti is provided by the shape of the straw-wrapped bottle. What a proper blend of the wines of these three grapes would yield in California remains to be discovered by some enterprising grower. San Gioveto requires cane-and-spur pruning to yield good crops.

Souzao, Tinta Cao, Touriga. Regions IV and V. Three classical varieties from the upper Douro Valley of Portugal, the region true port comes from. They do well in California's hot country, not only for their special purpose but for red table wine superior to the varieties commonly grown in the Central Valley.

**Tannat.* Regions I, II, and III. Midseason. Vigorous vine, healthy, production above average. The Tannat, a vine which is grown mainly in the southwest of France, on the foothills of the Pyrenees, is a variety of great age and respectability, source (when blended with Cabernet) of the famous Vin de Madiran of the Middle Ages. In any rational hierarchy of vines, this one falls just below such aristocrats as Cabernet and Pinot Noir. In California it is not grown very widely yet. But in the cooler regions it has been found to produce excellent wine of good body, deep color, a highly characteristic bouquet, and full flavor. Its wines are progressively heavier and less distinctive when grown in the warmer regions. Even so, they are superior in quality to the wines of the "standard" California varieties in the hot regions. Cane-and-spur pruning.

**Tinta Madeira.* Regions IV and V. Vigorous vine, above average productivity with spur or short-cane pruning. This is one of the numerous Portuguese varieties used for the production of fine port. It is in no sense a dry-wine grape, and its use in California is properly confined to the hot districts, for its fruit has high sugar and rather more acidity than most of the other Port varieties — a desirable characteristic. Tinta Madeira port is one of the best of those now being produced in California, being rich, smooth, and delicate.

**Trousseau.* Regions III, IV, and V. Early midseason, vigorous, productive, growing always with a heavy mantle of foliage which protects the fruit from sunburn. Like the Tinta Madeira, this is a Port grape, and is evidently identical with the variety known in Portugal as Bastardo. Though in certain parts of France it is also used for dry red wine, under the hot conditions prevailing in California its only proper use is for making port. By itself, the resulting wine lacks color (being what the Port producers call tawny) and requires a good deal of care in the making, owing to its very low acidity. Nevertheless, the so-called Trousseau Ports have been much admired in the past, and will continue to be. It seems to give best results in Regions III and IV. Spur pruning.

**Valdepenas.* Region IV. Vigorous, highly productive variety from the Valdepenas region of Spain, a worth-while addition to California viticulture because of its ability to produce well-balanced red table wines, better than the average in quality, in the hottest districts. Spur pruning.

Zinfandel. Regions II, III, and IV. Vine of moderate vigor and somewhat susceptible to mildew, but highly productive. How the Zinfandel came to California, and where it came from, no one knows. Be that as it may, it is one of the jewels in California's viticultural crown, being very widely planted and the source of a wine, usually well above average in quality, which is grown in California alone. Grown in Region II, it is capable of producing wine of distinctly superior quality with a brilliant and attractive color, a delicious and peculiar "sweet" bouquet, rich body, and excellent flavor. Good Zinfandel is a standing reproach to those California wine-makers who suppose that wines must be dressed up with French names in order to gain a market. In the warmer regions it becomes progressively heavier, less distinctive, and flat in flavor. Yet even in the heat of Region IV it is still capable of making tank car wine of satisfactory quality. The only reason for failing to recommend it unreservedly is that it is already well known and widely planted.

WHITE-WINE VARIETIES

Aligoté. Regions II and III. Early midseason. Vigorous vine, above average productivity. The Aligoté is what might be called the second-string grape, for white wine, of the Champagne and Burgundy regions of France. Its wine is vastly inferior to that of Chardonnay, but it is more productive; and, by comparison with most of California's present white wines, may be rated as good.

Burger. Regions II and III. Midseason, vigorous vine, highly productive. The Burger is widely planted in California chiefly for its high production and the supposed neutral quality of its wine, which (so the theory goes) makes it suitable for "stretching" other wines of superior quality. Actually, Burger wine is coarse, heavy, and flat-tasting, and that let-down feeling which the wine-drinker so frequently experiences when he tries a run-of-mine bottle of California white wine is all too frequently traceable to this variety.

**Chardonnay*. Regions I and II. Early. Vigorous of vine, not highly productive. The fruit is borne in small clusters, well filled but not compact, and for satisfactory crops the vine therefore requires cane-and-spur pruning. It is not greatly troubled by disease, and comes easily to maturity in the regions mentioned, with an almost ideal balance of sugar and acid. Chardonnay is one of the handful of vines yielding wine of the very highest quality. It is the source of those magnificent white Burgundies, Meursault and Montrachet, and of the famous white wines, of almost comparable quality, which are grown in the Beaujolais district and have the name of Pouilly. The wine has a highly characteristic bouquet and a delicious "stony" flavor. Fortunately, this vine is very well adapted to the cooler regions of California, where its wine is entirely typical. Beyond question, the Chardonnay makes the best dry white wines of California. In Region III, if picked promptly, it likewise makes superior wine, though of somewhat heavier body.

Chenin Blanc. Regions I, II, and III. Vigorous vine, above average productivity, sufficiently resistant to disease. This variety is the source of the delightful white wines of the Loire Valley in France, of which Vouvray and Saumur are the most familiar — wines that are fresh, light, fruity, sometimes slightly sparkling, sometimes slightly sweet. One of their most delightful characteristics is that you can never be sure in advance just what their character will be. In California they carry over only traces of their typical quality; and there, in Regions II and III, their best prospect is for the production of natural sweet wines more or less on the order of Sauternes. However, this variety is well worth trying for wines of more typical character in the coolest locations.

**Colombard*. Regions III and IV. Midseason. Vigorous vine, heavy and reliable producer. A grape from the Charente, or Cognac, region of France, grown there for the same reason as Folle Blanche — namely, its good acidity and the suitability of its wine for distilling into brandy. In California (where it has two synonyms: West's Prolific and Winkler) its utility is obvious. It is one of the few grapes that develop sufficient acidity for a good dry wine in the hotter areas. In character the wine is neutral and clean, pleasant and refreshing. It is useful also for blending with less acid varieties. Spur pruning.

Emerald Riesling. Regions II and III. The true Riesling does not give fully characteristic wines in California. This is a hybrid, or *métis,* intended to combine the Riesling aroma with better acidity in hot climates. Though better adapted to California heat, it is still not equal to the real thing grown in the right place, which is to say the Rhine and Moselle valleys.

**Folle Blanche.* Regions II and III. Midseason. Vigorous vine, average productivity. Fruit is borne in very compact clusters, which are highly susceptible to rot under humid conditions. This is the classic grape of the Charente district of France, where it is grown as the basis of the finest Cognacs. For the best brandy, wine of low alcoholic content and high acidity are desirable. In the Cognac district the Folle blanche wine is of this character, being too acid for drinking and too low in alcohol for stability. It is likewise grown under the name of Piquepoule, in the South of France, where sufficient acidity is always a problem. In the Midi its wine has correct acidity, sufficient alcohol, and a good deal of body. It is suitable for a table wine of average quality and as a base for fine Vermouth. In California, in the regions mentioned, it produces wine very much like that which it produces in the South of France. It has a pleasant aroma, clean flavor, and a tartness all too rare in California white wine. The Californians use it, with dubious success, as a base for their sparkling wines. Its brandy possibilities are being investigated. Spur pruning.

Gray Riesling. Regions I and V. Not to be confused with the true Riesling. This is a vigorous variety, ripening in early mideason, requiring cane-and-spur pruning for adequate crops, and rather susceptible to mildew. Quality of its dry white wine, though perhaps better than the average, does not place it in the rank with the true Riesling. But it is good for sweet dessert wines when grown in Region V, the hottest of California's viticultural regions.

Marsanne. Vigorous and productive. It is a vine from the Rhone Valley; and, theoretically at least, the vines that flourish on the sun-baked banks of that historic stream should do beautifully in California. Certainly the similarity of climate produces similarity of results in the case of Petite Sirah and a number of other varieties. However, Amerine and Winkler find that the white wine of Marsanne is coarse, badly balanced, and generally inferior to a number of other white-wine varieties, even including Burger. Hence they do not recommend it. However, in the Rhone Valley this variety finds its best place not for white wine but in the blend which produces such very fine red wines as Châteauneuf du Pape and Hermitage. The wine-growers of the Rhone Valley are an enlightened folk, and if Marsanne failed to contribute anything of value to these blends they would long since have abandoned it. It is certainly worth trying for a similar purpose in Regions II and III.

Muscadelle de Bordelais. Region III. Vigorous, and sufficiently productive with cane-and-spur pruning. This is the third, and least important, of the three grapes used in France for the production of Sauternes. Its role is to supply a certain element in the Sauternes bouquet. It is a muscat variety, and like all muscats it does not yield an agreeable dry wine. Its sole purpose in California (and it isn't actually necessary for this) is in blending to produce wines of the Sauternes type.

Muscat of Alexandria. Rather weak vine, heavy producer. This variety, probably more widely grown than any other grape in the world, is mentioned here only as a warning. It is a "triple threat" grape as grown in California, being used chiefly for the production of muscat raisins and somewhat for shipping as a table grape. The inevitable surplus always goes to the wineries for the production of the cheapest heavy sherries, muscatels, and so on. No one seeking wine of quality should go near this grape.

**Muscat Canelli.* Regions III and IV. Vine of moderate vigor and moderate productivity, ripening in late midseason and yielding rather light crops. Another muscat variety, source in southern France of the famous sweet and liquorous Muscat de Frontignan, and also used in Italy in the making of the sparkling muscat wine called Asti spumante. In short, a superior grape for those interested in producing a sweet muscatel of the best character. Spur pruning.

Orange Muscat. Regions IV and V. Midseason. Vigorous vine, and a heavy and reliable producer of large clusters of berries of uneven size. This muscat variety is relatively little planted in California at present. Muscatel made from it is far superior to that made from Muscat of Alexandria, being remarkably rich and aromatic, yet not disagreeably so. It is adapted, of course, only to the hot districts. Spur pruning.

**Palomino.* Regions IV and V. Early midseason. A vine of great vigor and high productivity, relatively immune to disease difficulties. In California it is very widely planted, and on it (and the Burger) we may lay much of the responsibility for that coarseness of flavor which characterizes so many of the California white wines. Its dry wines have an aroma resembling an inferior apple cider in which pears have somehow been mingled. It is more frequently encountered under the misnomer Napa Golden Chasselas than under its true name Palomino. This grape, despite its unsuitability for dry white wines, nevertheless has great virtues. As its name suggests, it is of Spanish origin; and in Spain it is one of the two basic grapes for sherry-making. At present the best of the California sherries are being made of this variety; and of these, the best will in the future doubtless come from the hotter districts, where Palomino, as a sherry grape rather than a dry-wine grape, reaches perfection.

Pedro Ximenes. Probably the most famous of the several Spanish varieties used for the production of sherry. In California the growers have shown a curiously perverse tendency to use almost *any* varieties for sherry except the varieties used for that purpose in the vineyards around Jerez. Recently they have begun to swing over. Palomino, as was suggested above, is now being used for its proper purpose, which is sherry-making. Pedro Ximenes, which is an even better sherry grape than Palomino, deserves wider trial.

**Peverella.* Regions III and IV. Vigorous, healthy, above-average producer. This grape, grown a good deal on the southern slopes of the Alps, is one of the few white-wine grapes capable of producing a white wine that is light and refreshing in the hot districts. Even so, it falls down on the job when planted in Region V. Its wine, though not fine, is above average. Spur pruning.

***Pinot Blanc*. Regions I, II, and III. Early. Vine of below-average vigor, yielding relatively small crops of tight little bunches that are rather susceptible to rot when rains coincide with the ripening season. The Pinot Blanc has in the past been confused a good deal with the Chardonnay, since both are grown in the fine-wine districts of eastern France and since their fruit yields wine of quite comparable character. Like those of Chardonnay, wines of Pinot Blanc have the highly characteristic aroma associated with white Burgundy, plus a "stony" flavor. They retain these characteristics fully when grown under the proper conditions in California, which are found in Region I (the coolest region) and in the cooler parts of Regions II and III. In hot seasons the grapes of this variety tend to lack acidity, and their wines to be heavy and without marked character. Spur or short-cane pruning.

Piquepoule. See Folle Blanche.

Red Traminer. Early. Region I. A grape of moderate vigor producing small but regular crops. This is one of the true fine-wine varieties, and is called Gewurz Traminer in the wine-growing parts of Germany, and Savagnin in those parts of France, namely, Alsace and the Jura Mountains, where it is grown. A fine Traminer wine is fit to be compared with the very best Riesling wines of the Rhine and Moselle valleys; and there are those who consider it a superior grape to the Riesling. Its wine is indescribably delicate, with an aroma like no other. In California, with its warm and even climate, it is not at its best, however; only in the coolest parts does it show a proper balance of constituents at maturity. But the search for a proper location for this grape is well worth the effort. Cane-and-spur pruning.

***Sauvignon Blanc*. Regions I, II, and III. Early. A vine

of moderate vigor, yielding crops of below-average size, owing to the fact that the clusters are small. It should be spur-and-cane-pruned in order to obtain an adequate yield. This vine is one of the three varieties (the other two being Sémillon and Muscadelle de Bordelais) responsible for the true Sauternes, and constitutes approximately one-sixth of the traditional blend. Its fruit has a delicious, and very special, aroma and flavor, and develops high sugar content. Its role in the Sauternes blend, in California as well as in France, is to contribute aroma and sweetness; and the resulting wine has a right to be placed beside that of Chardonnay as one of the two best white-wine types that California is capable of producing. The best California sauternes, made with the help of Sauvignon Blanc, is an exquisite wine, rich, soft, perfumy, and strictly comparable to French Sauternes. Sauvignon Blanc is also used in parts of France for production of an aromatic dry white wine; and by some is very much appreciated, though it usually has an undertone of bitterness in its flavor. Here it is also used by itself or with Sémillon for a dry white wine, in which role it often produces truly distinguished results.

**Sémillon.* Regions II and III. Midseason, vigorous vine, production well above average, fruit subject to cracking and rotting when rain coincides with ripening. This is the principal grape in the traditional French Sauternes blend, accounting for two-thirds of the whole, and it is also the informing grape in those somewhat similar French wines Graves and Monbazillac.[1] It is primarily, then, a grape for the production of natural sweet wines, and it reaches the peak of its quality in California in Region III and the hotter parts of Region II. In the cooler regions, it is also used a good deal for the production of dry white wine, but its aroma and flavor are less congenial to a dry wine. Better, probably, to use it only for its best purpose, the production of natural sweet wine. Since it is low in acid, it must be fermented with care. Spur pruning.

[1] Graves is usually less sweet than Sauternes, Monbazillac usually even sweeter. All three have a similar bouquet.

**Sylvaner*. Region II. Early. Vigorous vine, rather subject to mildew and rot, producing average crops. The Sylvaner is the second-string Rhine-wine grape. In the valleys of the Rhine and the Moselle, and in Alsace, it is grown in the second-best locations, where the Riesling does not produce wine of the highest quality. Its wine has a certain distinction of character and a good balance of sugar and acidity, and is usually drunk young. The carafe wines served in the inns and hotels of that part of Europe are usually Sylvaner. In California its wines are distinctly above average in quality, being less heavy-bodied and more delicate in constitution than most of the other white wines – though still falling short of its European character. Its best wine is produced in Region I, but there the mildew and possibility of rot are serious handicaps, so that, if it is to yield sure crops, its proper place is the warmer Region II. But there, if the wine is not to be just another heavy-bodied, rather coarse wine, close attention must be paid to the picking time. Spur pruning.

Ugni Blanc. A reference to this grape is inserted only because it is already grown to a limited extent in California and is marketed under this name. This is one of the most ancient of white-wine varieties, and is widely dispersed throughout the world. Under the name of Trebbiano it is grown a good deal in Italy, and has been since Roman antiquity. Under the name of St. Emilion, it is grown in the valley of the Charente for the production of cognac. Its wine is light, neutral, not unpleasant, but without any special virtue.

**Verdelho*. Region IV. Early. Vigorous vine, and highly resistant to disease. However, owing to the smallness of its bunches, it yields only moderate crops even when spur-and-cane-pruned. The Verdelho is a Madeira grape; and those who have long felt that the so-called sherries of California more closely resemble the wines of Madeira than they do true sherry (because of the California practice of "cooking" the sherry wines in the course of aging them) will not be surprised that this grape, under California conditions, gives a dessert wine of distinctly superior quality. It will probably never be widely planted because of its relatively low production; but for the experimental it is distinctly worth a trial.

White Riesling. Regions I and II. Early. Spur-and-cane pruning. Vine of only moderate vigor, producing moderate crops of smallish bunches. This variety, being well adapted to the humid conditions of central Europe, is sufficiently resistant to mildew and the other diseases of fruit and vine. The Riesling ranks as one of the world's two or three finest white-wine grapes. It is responsible for the reputation of all the most famous wines of the Rhine and Moselle valleys, produces most of the fine white Alsatian wine, and is widely spread throughout central Europe, everywhere (save as the Mediterranean is approached) producing wine of superior quality. The fruit has an unmistakable aroma and flavor, the source of the characteristic and likewise unmistakable aroma of the wine. Unfortunately, the Mediterranean climate of California is not ideal for this grape. The fruit tends to overripen and is usually deficient in acidity. For best results, it requires the relatively cool conditions of Region I. And even there, though its wine is above the average, it is by no means comparable in delicacy, balance, and general finish to the Riesling wines of Europe.

A NOTE ON ROOTSTOCKS

The necessity for grafting *vinifera* vines if they are to escape injury from the phylloxera has been explained. There are three other good reasons for grafting in special circumstances. One is to increase the vigor of weak-growing varieties by putting them on strong-growing stocks. Another is to induce early ripening of the vine's wood and so improve its winter hardiness. The third is to protect the roots of susceptible varieties from injury by root rot and nematodes in parts of the United States where these hazards exist.

These are the first considerations in determining choice of rootstocks. But there are several other factors to be considered. Rootstocks must take grafts easily and they should root easily. There must be good affinity with the scion that is to produce the crop. And, finally, the stock must be suitable for the soil where it is to be planted.

As to this latter point, there are several general guides. The species *rupestris* roots deeply: hence stocks containing it are normally preferable in hot, dry conditions where the soil is deep; but it dislikes lime. The species *riparia* has a shallow rooting system: hence stocks containing it are normally preferable where soil is shallow or water table is fairly high. It prefers soil on the acid side. The species *Berlandieri* tolerates more lime than most American species: hence it is the species of choice where the soil is on the alkaline side of neutral. The species *Champini* and *Lincecumii* have shown resistance to nematodes and certain forms of root rot: their use is indicated where such troubles are likely.

Aramon x *rupestris Ganzin No. 1*, called A x R No. 1 for short, is widely used in California on deep, low-lime soils.

Mourvèdre x *Rupestris 1202*, or simply 1202, has somewhat more lime tolerance than the above, and somewhat more phylloxera resistance, and does better on shallow soils.

Rupestris St. George is the California stand-by. It is well adapted to most viticultural parts of the state, highly resistant to dry conditions and to phylloxera, and a vine of great vigor. It is so vigorous indeed that it promotes coulure on some *vinifera* scion varieties. This can be counteracted to some extent by pruning more generously for a larger crop.

Berlandieri x *Rupestris 99-R.* On form, the combination of these two species should have wide use in California, and tests so far show good promise. This is the best of the so-called Richter rootstocks.

Solonis x *Othello 1613.* Because of its combined resistance to nematodes and phylloxera, this has been found a useful stock in California where nematodes are troublesome.

Dogridge and Salt Creek. Two seedlings of *V. Champini* which are resistant to nematodes, are extremely vigorous, and have shown themselves to be widely adapted in California and in the Southwest, where *Champini* comes from.

As might be expected, the rootstocks proving most useful in the East are not the same as those that are best for California. Some of the ones being tried in the East are:

Teleki 5BB (Berlandieri x *riparia).* One of the most generally adaptable. It does well on loam and clay, on "cold" and wet soils. Much used in northern Europe.

SO_4 (Berlandieri x *riparia).* Similar to 5BB, but ripens sooner and has somewhat less vigor, factors that help it to ripen the wood of the scion thoroughly.

Riparia x *Rupestris 3306 and 3309.* Two of the oldest French rootstocks still much used in northern European wine-growing areas and found useful in the Niagara peninsula of Canada and in New York. The more vigorous of the two is 3309, which has also indicated some nematode resistance.

Solonis x *Riparia 1616.* High resistance to phylloxera. Behaves very well on gravelly loam, also on wet, heavy, quite acid loam. It has shown excellent affinity with the early *vinifera* currently under trial here and there in the East. In California it has also proved good on sandy loams. The parent *Solonis* is itself a complex vine, offspring of a cross between still another Texas species, *V. candicans*, and a *riparia-rupestris* hybrid. *Solonis* is still used a good deal in the French Camargue, the swampy land just west of the mouths of the Rhone that empty into the Mediterranean, because of its tolerance to wet soils with fairly high salt content.

Chapter XIII

GRAPES FOR THE EAST

In the course of this book I have often emphasized the dual character of American viticulture: on the one hand, the viticulture of California (with offshoots to the North in Oregon and the Southeast in the Rio Grande Valley); on the other; the viticulture of the rest of the continent, which save for a few favored localities is uncongenial to the *vinifera* because of climate and the prevalence of vine diseases. In this much larger and highly variegated area, only the wild vines and their hybrids are really at home. The "East," as used in this chapter, includes areas as different as Washington, Arizona, Florida, and Massachusetts.

Most of these hybrids contain a proportion of *vinifera* parentage. If they are first-generation hybrids, then of course they are 50% *vinifera.* Thus the old American hybrid Isabella had one pure *vinifera* parent and one pure *labrusca,* or fox grape, parent. The French hybrid Baco No. 1 had one pure *vinifera* parent and one pure *riparia* parent. Or the hybrid may be considerably more complex. Thus, another American hybrid Ripley had Winchell and Diamond as its parents—both of these in turn being one-half *vinifera* and one-half *labrusca.* So Ripley is also 50% *vinifera* but got its *vinifera* through two distinct lines.

Or a hybrid may be much more complicated than that, as is the case with most of the French hybrids, which are the result of many crossings and recrossings, or of the native variety Delicatessen, which contains the parentage of no fewer than four species, *Lincecumii, labrusca, vinifera,* and *Bourquiniana.* In the continuing work of hybridization, parentage becomes more and more confused and diffused, until it loses much of its meaning. Hybrids of later generations become mixtures of genes carrying specific characteristics rather than of species, one gene or group of genes contributing winter hardiness, another earliness, another color, another desirable flavor, and so on.

For most people it is sufficient to say that in general the *vinifera* parentage accounts for high quality of fruit: handsome appearance, high sugar content, rich and delicate flavor tending toward neutral, and high productivity. By contrast the factors provided by the wild American species are those closer to nature—vigor of vine, resistance to disease and adaptability to American climates, and frequently self-sterility and fruit in which the emphasis (as always in nature) is on procreation[1] rather than the factors which makes grapes attractive to man.

The United States is rich in native species of *Vitis*, but relatively few of them have proved to be profitable material for hybridization. These few are shown on the table on pages 200–1.

Of these species, fewer still have so far yielded results in the form of new hybrids of practical value for wine-makers.

Before getting on to descriptions, let us distinguish several rather different groups of hybrids.

There are the old familiar American varieties which were discussed in Chapter II. Most of these were crosses that occurred spontaneously in nature and were spotted by sharp-eyed amateurs and so found their way into cultivation. Thousands of these natural or accidental hybrids have been grown, named, and described. Most of them are now ghosts—they had a brief period of culture and temporary local popularity and then passed from the scene. Of those still to be found, the greater number are without value as wine grapes, either because of cultural defects or because of obvious deficiencies of the fruit. Even after a preliminary screening, most of those remaining have never had a really

[1] That is, small berries, lots of seeds, and thick skin to protect them.

GRAPE SPECIES USED IN BREEDING HYBRIDS

(From the *Yearbook* of the United States Department of Agriculture, 1937)

SPECIES, COMMON NAME, AND NATURAL RANGE	CHARACTER OF VINE	RESISTANCE TO					QUALITIES FOR BREEDING
		Phylloxera*	Cold**	Heat**	Wet**	Dry**	
Vitis aestivalis Michx.; summer grape. New England to Georgia and westward to the Mississippi River.	Vigorous, climbing; leaves large, 20 cm., 3- to 5-lobed.	14	VG	G	F	G	Resistance to fungus diseases; high sugar percentage; suitable wine properties; possible table use if crossed with large-berried varieties.
V. aestivalis var. *bourquiniana* Bailey (*V. bourquiniana* Muns.); Bourquin grape. Origin doubtful; adapted to southeastern States.	Vigorous, climbing; leaves large, 3- to 5-lobed.	—	F	G	F	G	Vigor; disease resistance; productiveness; colored juice.
V. Berlandieri Planch.; Spanish grape, winter grape. Texas and northern Mexico.	Medium vigor, slender; leaves medium, 10 cm., 3- to 5-lobed.	19	F	G	F	G	Rootstock resistance to phylloxera; ability to grow on strong, limy soils.
V. candicans Engelm.; the mustang grape. Mainly Texas, parts of Arkansas, Oklahoma, Louisiana, and Mexico.	Very vigorous, high climbing; leaves medium, nonlobed to 3-lobed.	15	F	G	F	G	Vigor for rootstock; is easily hybridized; adapted to black limestone lands; large-berried fruit for wild vine.
V. Champini Planch.; the Champin grape. Mainly Texas.	Very vigorous, climbing; leaves medium, 10–12 cm., nonlobed to 3-lobed.	15	F	G	G	G	Vigor for rootstock; healthy foliage; wide adaptability; large-berried fruit.
V. cordifolia Lam.; frost grape. Wide range, from Great Lakes to Florida.	Vigorous, climbing; leaves medium, 10 cm.	18	G	G	G	G	Vigor; phylloxera resistance; wide natural range; but poor fruit.
V. labrusca L.; fox grape. New England to northern Georgia, westward to Indiana, and bordering the Ohio River.	Medium vigor, climbing; leaves large, nonlobed to slightly lobed.	5	VG	F	F	F	Cold resistance; large-berried fruit; strong, distinctive flavor.

* Phylloxera resistance graded 1 to 20.

** Symbols for resistance to cold, heat, wet soil, drought: VG very good; G good; F fair.

GRAPE SPECIES USED IN BREEDING HYBRIDS

(*Continued*)

SPECIES, COMMON NAME, AND NATURAL RANGE	CHARACTER OF VINE	RESISTANCE TO					QUALITIES FOR BREEDING
		Phylloxera*	Cold**	Heat**	Wet**	Dry**	
V. Lincecumii Buckl.; pinewoods grape, post-oak grape. Texas, parts of Louisiana, Oklahoma, Arkansas, and Missouri.	Vigorous, bushy to climbing; leaves very large, 3- to 5-lobed.	14	F	G	G	VG	Vigor; disease resistance; large clusters and berries; strong flavor.
V. Longii Prince; Long's grape, bush grape. Parts of Arkansas, Oklahoma, Texas, New Mexico, and southeastern Colorado.	Very vigorous, bushy to climbing; leaves large, 3- to slightly 5-lobed.	14	G	G	VG	G	Vigor; phylloxera resistance; easy rooting of cuttings; vinous flavor.
V. monticola Buckl.; sweet mountain grape. Texas.	Medium vigor, slender, climbing; leaves small, nonlobed to slightly 3-lobed.	18	G	F	F	G	Phylloxera resistance; health of foliage; fruit medium to small.
V. rotundifolia Michx.; muscadine grape. Potomac River to Florida and west to eastern Texas.	Vigorous, slender, climbing; leaves small, not lobed.	20	F	G	F	G	Disease-resistant vine and fruit; special fruit flavor.
V. rupestris Scheele; sand grape. Southern Missouri and Illinois, Kentucky, Tennessee, Oklahoma, and eastern and central Texas to the Rio Grande.	Very vigorous, bushy, rarely climbing; leaves small, mostly nonlobed.	19	F	G	G	G	Phylloxera resistance; easy propagation; vigorous.
V. riparia Michx.; river-bank grape. Canada to Texas and west to Great Salt Lake; wide range.	Vigorous, slender, moderately climbing; leaves large, mostly nonlobed to slightly 3-lobed.	19	VG	F	G	F	Phylloxera resistance; cold resistance; easy propagation.
V. vinifera L.; European grape, wine grape. Introduced species.	Medium to strong vigor, bushy to climbing; leaves mostly 3- to 5-lobed, occasionally 7-lobed.	1	F	VG	F	G	Productiveness; high quality; easy propagation; some seedlessness.

* Phylloxera resistance graded 1 to 20.

** Symbols for resistance to cold, heat, wet soil, drought: VG very good; G good; F fair.

critical test of their value for wine. During the latter part of the nineteenth century, when the wine-making possibilities of our native grapes were being eagerly explored, a great many earned unmerited praise merely because they were better than no wine grapes at all. Enthusiasts, finding that wine of a sort could be made from these, praised them in a thoroughly uncritical manner. Over and over again in nineteenth-century writings one finds varieties of indifferent quality being praised as capable of producing "wine comparable to the finest burgundy or claret" or "wine far more delicious than the best rhenish wines and moselles." The enthusiasm of these pioneers was understandable, but in the light of subsequent experience we are not compelled to share it.

Then there is a large group of man-made American varieties, most of which are also now ghostly. If vines mate spontaneously in nature, why not plan and carry out artificial matings by cross-pollinating selected vines? Captain Jacob Moore of Concord, Massachusetts, and Edward S. Rogers of nearby Salem, were conspicuous among nineteenth-century experimenters, and a few of their productions linger on. But they worked exclusively with the *labrusca* species, and though their grapes were good to eat they had too much foxiness for wine.

Among the Americans who worked with American species, the most lasting contribution was made by T. V. Munson. He worked in the Southwest, in the general area of the Red River Valley, where *labrusca*, ubiquitous in the Northeast, gives way in nature to such non-foxy species as *rupestris*, *Lincecumii*, *Berlandieri*, *monticola*, and others. His productions represented a sharp break with the *labrusca* hybrids of the Northeast, and provided dazzling glimpses of the possibilities of hybridizing for wine production, even though few of them survive today. One trouble is that Munson's first interest was the breeding of table grapes. Though in his writings he referred often to the wine-making possibilities of his hybrids, he himself confessed to a very tepid interest in wine. Only a few of his dozens of promising varieties have even been used for wine and subjected to the critical examination all French hybrids get.

A third group consists of the more recent productions of a few of our experiment stations, notably the New York Agricultural Experiment Station at Geneva, New York. Here too the emphasis has been mainly on table grapes. A few show promise as wine grapes, and there has lately been some better-focused activity in wine-grape breeding. But as of now none of the New York hybrids has caught on in a serious way.

Finally there are the French hybrids, whose potential value for American viticulture was discussed at some length in Chapter III. These are called French hybrids because they were produced by Frenchmen working in French vineyards. But the materials they worked with, and are still working with, belong to the same pool of plant material —the various American species and the *vinifera* grapes of Europe. The offspring of these matings can as well be called Franco-American hybrids, or American hybrids for that matter; and in fact one current name for them in France is *vignes Américaines*. The difference is that the production of these has been a much less hit-or-miss affair than grape breeding in this country, having been undertaken for the most part by men with a clear idea of what they wanted and a pretty good notion of how to get it. Applied science is not too flattering a term for their work—though several of the most successful French hybridizers were not in fact men of scientific training.

Hence, for the most part, a systematic rather than a romantic or fanciful nomenclature (involving the names of places, old friends, wives, sweethearts, favorite daughters, legendary characters, and so on) was more appropriate. With few exceptions, the French hybrids are known merely by the name of the hybridizer, with the original seedling number added. For anyone not familiar with the vines themselves, these names make singularly dull and uninspiring reading. But there is all the difference in the world between, let us say, Seibel 5279 and Seyve-Villard 12375, or between

Joannes-Seyve 26-205 and Vidal 256. For those who have studied the vines in question, these numbers are a very special poetic idiom evoking distinct plant personalities, each with its faults and crotchets, singular beauties, and particular virtues. The numbers also summon, for the knowing, the recollection of wines as different from one another as wines can be, different in ways which the poor vocabulary of odors and flavors is quite unable to communicate.

In making a list of the French hybrids, the first desideratum had to be the vine's actual presence in vineyards somewhere in the United States. This placed an automatic limit on the list. Yet at the same time it had an advantage, since by and large the hybrids now domiciled here represent the cream of the crop. Some that are still grown to a considerable extent in France but are on the way out, such as Gaillard-Girard 157, are absent here. Nor do we have the annual plague of new and untested varieties, many of them tempting in the beginning but likely to reveal defects later on. Yet, even of the vines available, only those which have already proved themselves or which show definite promise should be described. And of those described, almost all have been tested at Boordy Vineyard for varying lengths of time. This is not a bookish compilation but an outgrowth of personal evaluation.

Relatively few hybridizers are represented on the list. Maurice Baco, who did his work in the Landes—that viticulturally difficult region south of Bordeaux—is represented by two only. Couderc, one of the earliest and one of the greatest, is represented by two. His pioneering work was valuable chiefly in providing parents for later hybridizers. There are several Ravats. The greatest number is from the Seibel collection. Louis Seibel was a prodigious and untiring if not always critical worker, and his vines offer a large array. The name Seyve-Villard is represented by several remarkable vines, of which S.V. 12375 is the most largely planted hybrid in France.

The names of many of the pioneers are missing: Malègue, Gaillard, Castel, Oberlin, Humbert et Chapon, Roy-Chevrier, Chevalier, Péage, Ganzin. Their productions have gone the way of the early American hybrids of Moore, Rogers, Ricketts, and the rest. There are the names of contemporaries: Burdin, Galibert, Kuhlmann, Joannes-Seyve, Landot, Vidal.

The period of serious testing of these vines has been short, as such work goes—a few decades only. Although some of them are so far along that they have entered large-scale commercial viticulture, most are not. The labor of adjusting the right varieties to the right places is slow, but fascinating. Mistakes have been and will be made. Because so much has still to be learned about their behavior under American conditions, judgments at this point are inevitably tentative.

So we come to the descriptive list of grapes for the East: first the American hybrids, or native varieties, for white wine and red wine; then the French hybrids for white wine and red wine. Except when otherwise stated, cane-and-spur pruning may be assumed. Length and number of canes must be left to the judgment of the grower, being determined by the vigor and productivity of the vine.

NATIVE VARIETIES—WHITE WINE

Because the possibilities of our viticulture have been so greatly enlarged by the French hybrids, marginal varieties have been eliminated here—the maybes, the ifs, and the buts of our viticultural heritage. The following varieties are currently in substantial production or have permanent value.[1]

Catawba. (Lab., vin.) Midseason. Districts 2;3; cooler and less humid parts of 4 when grafted. This is the most famous of the native white-wine hybrids, a natural seedling. Vine is fairly vigorous and productive, but neither very

[1] Varieties described in the 1945 edition of this book but now eliminated are Adams Champagne, Augustina, Brocton, Diana, Dunkirk, Golden Muscat, Hector, Iona, Ontario, Peabody, Ripley.

hardy nor very resistant, being subject to mildew of foliage and black-rot of fruit. Moderate cane-and-spur pruning. Fruit is not particularly agreeable to eat, and the must has a pronounced foxy aroma and high acidity coupled with fairly high sugar content. Its still wine is dry to the point of austerity, and has a very clean flavor and a curious, special, spicy aroma. Its main use, however, is as a blend for sparkling wine, since it takes the secondary fermentation very cleanly, always falls bright, is stable, and does not tend to darken with age. Though there are better wine-grapes, it has a definite place in American viticulture and an assured future, if only because a century of use has taught the winemakers how to handle it.

Delaware. (Bour., lab., vin.) Early. Districts 2;3;4 (when grafted); warmer parts of 6 and 8; cooler parts of 7. Usually considered the best native white-wine grape. The vine is far from vigorous, and is highly susceptible to mildew. Nevertheless, the mildew is easily controlled by spraying; the vine is remarkably winter-hardy, and it has an extremely high range of adaptability as to soil. Productivity modest, owing to the small size of the bunches. Fruit is borne in small, very handsome clusters, usually shouldered, of pretty pink grapes. By planting closer than the average in rows, and pruning to long canes with spurs, a good production per acre can be attained. The wine is light golden in color, characterized by a rather pronounced and delicious non-foxy aroma, softness of flavor, and considerable body in certain regions. It is ordinarily used in the American champagne blends, in combination with Catawba, and for this purpose brings prices so high that it is practically excluded as a source of still wine. The wine has a tendency to darken with age, and to oxidize unless it is well cared for.

Dutchess. (Vin., lab., Bour.) Early midseason. District 2. An old New England hybrid which would be more widely planted except for the weaknesses of its vine. Vine is not vigorous, does not like rich or heavy soils, and is subject to the fungus diseases. Fruit also has a tendency to crack when ripe unless picked promptly. Moderately productive. Clusters of average size, very handsome, pale yellow, compact, shouldered, delicious to eat. Wine is agreeable, slightly foxy, of good quality.

Ellen Scott. (Linc., lab., vin.) Districts 3; drier parts of 4; much of 7. Midseason. A vine of great vigor, good health, and high productivity. Fruit is remarkably handsome, the clusters being very large, well shaped, and evenly colored. Berries range in color from very dark lavender to rose, the color factor evidently being affected by soil, or climate, or both. Flavor is mild and attractive, sugar content running from 18–20% along with good acid balance. Has been recommended frequently as a wine-grape, but I cannot testify as to the quality of its wine. A Munson hybrid, named for Mrs. Munson.

Elvira (Rip., lab.) Early. Districts 1 (?); 2; 3; possibly parts of 6 and 8. An old *riparia* hybrid, with vigor, good health, perfect winter-hardiness, and above-average productivity. Clusters rather small, so that long-cane pruning is necessary for an ample crop. Berries green to gold, with a tendency to crack in wet weather. The must is rather low in sugar and high in acid; and the resulting wine has sometimes a rather sharp aroma and a "hard" flavor, but stable constitution. Yet it can be very agreeable. This grape is used for still wine and also in the American sparkling-wine blends.

Herbemont. (Bour.) Late. Districts 4 (least humid parts); 7. One of the oldest of the American white-wine varieties, of mysterious origin, and one of the few American hybrids planted to any extent abroad. In soil that suits it (that is, moderately rich and well drained) it has vigor and productiveness, but the foliage and fruit are both subject to mildew and blackrot; and the fruit itself rots when ripening coincides with high humidity. Clusters large and handsome. Berries below average in size, and variable in color, ranging from deep lavender to mouse gray, depending on local influences. Ample sugar and good acid balance. The wine is rather ordinary in quality. Its cultural defects are a serious handicap.

Isabella. (Lab., vin.) Early midseason. A good deal has been said of this variety already, in Chapter III. Vine is entirely *labrusca* in character, vigorous, sprawling and healthy. Productivity above average, large well-formed clusters of blue fruit, which is not heavily supplied with pigment. Fruit is highly foxy but, when pressed and fermented white-wine fashion, the resulting wine is pale, without pronounced foxiness, with the curious characteristic of growing paler with age. The New York State champagne-makers insist that this wine, in a blend, has the property of "bringing out" the attractive aroma of other grapes; but it is hard to resist a suspicion that this is a rationalization for Isabella's use. At the time of the phylloxera epidemic, the Isabella was widely dispersed throughout the vineyard districts of Europe, where it is now looked upon as being little better than a weed.

Missouri Riesling. (Rip., lab.) Districts 2;3; cool parts of 7. A white-wine variety grown considerably in the wine-making region around Hermann, Missouri, and to a limited extent elsewhere, at the turn of the century. It is vigorous and healthy, and moderately productive. It yields a highly aromatic white wine of fair quality, but is probably without interest for the future.

Niagara. (Lab., vin.) Early midseason. Districts 2, 3. Great vigor, with extremely high production of handsome fruit when grown in deep, rich soil. Its wine is extremely foxy. The variety has a vogue in the New York wine-making districts because of the popularity of a sweet white wine, called Lake Niagara, which is made from it. It has no great future as a source of dry white table wine.

Noah. (Rip., lab.) Midseason. Districts 2;3. A white *riparia* hybrid of the general class of Elvira, vigorous and fairly productive, bearing bunches of somewhat larger size and yielding a comparable wine. This is one of the grapes imported and widely planted by the French in the black days of the phylloxera epidemic, but now in disrepute among them because of the foxiness of its aroma. Following the repeal of Prohibition it had a temporary resurgence in the wine-making area of southern New Jersey, but it has no substantial future.

NATIVE VARIETIES—RED WINE

As we have seen, some of the native white-wine grapes have an assured future. Unfortunately there is not a single fully satisfactory red-wine native. The cause of this is the presence in almost all of them of some proportion of the foxy *labrusca* species. A white-wine grape such as Catawba is foxy to the taste. But the aromatic compounds which account for this are present mainly in the skin. White wines are made by pressing the fruit and fermenting the juice without the skins; most, though not all, of the foxy aroma is left behind with the skins. But red wines, being fermented on the skins, pick up all the foxiness (and other "wild" flavors contributed by certain other native species) in the course of fermentation. As wine consumers in this country become more critical, these flavors, together with the grapes which provide them, are bound to be less and less admired. This, then, is a very short list.[1]

Concord. (Lab. x?) The standard blue grape of the East now and for years to come, ripening in midseason, easy to protect against disease, fully hardy, vigorous, and productive. It is included here as a warning. It is delicious to eat fresh and as jelly, and makes tasty grape juice, and has long been used by eastern wineries as an all-purpose grape, mainly for want of something better: for ports and sherries, pressed fresh and decolorized for white wine, and for red table wine. Ways have been found of reducing its foxiness somewhat, and there will always be a market for its red wine among those who enjoy the vinous version of the foxy *labrusca* aroma. But those whose taste in wine is formed on European standards will not want to plant it. Nor should it ever be used in blends with the French hybrids. Its powerful aroma always predominates.

Cynthiana. (Aest., lab.[?]) Districts 3; cooler parts of

[1] Red-wine natives described in the 1945 edition but now eliminated are Alpha, America, Bailey, Buffalo, Clinton, Cloeta, Eumelan, Extra, Lomanto, Manito, Montefiore.

4;7. Late midseason. Until the appearance of the French hybrids, this was considered perhaps the best of the American red-wine hybrids, and is for all practical purposes the same grape as the variety called Norton, or Norton's Virginia, or Norton's Seedling. Though it originated in Virginia, it achieved its greatest popularity along the banks of the Missouri River, where at one time there was a substantial wine-making industry around the town of Hermann. Vine is vigorous and rampant in growth, but of less than average productivity owing to the small size of the clusters. Fruit regularly runs sugar above 20%, but this is coupled with excessive acidity, so that the must requires dilution. Even so, the wine has intense color, and a distinct, agreeable, non-foxy aroma. It can acquire bouquet with aging.

Delicatessen. (Linc., lab., vin., Bour.) Early midseason. Districts 3;7. Another Munson hybrid of complex parentage. Its vine is vigorous, with foliage that is dark, coarse, and resistant to fungus diseases. Average production, doing well under either Kniffin or cordon training. Its one serious cultural defect is a tendency of the berries to crack when high humidity coincides with ripening, and in consequence to develop rot. Clusters of above average size. Berries black, with practically no bloom, rather soft, and with intense color. The fruit has a rather pronounced and high characteristic aroma, moderate sugar, and acidity slightly above average. Sugar and acidity are *not* well balanced. Yet, properly ameliorated, this variety gives a red wine that is highly distinctive, with a special raspberrylike aroma and the unusual combination of deep color and rather light body. Actually its wine ought not to be compared to any of the conventional European kinds.

Ives. (Lab.) Midseason. Districts 2;3. Vigorous, hardy, productive. Fruit is strongly foxy, despite which it has been grown a good deal as a wine-grape. It shares with Concord the blame for the poor reputation of eastern red wines, and ought to be discarded.

Lenoir. (Bour.) Late. Districts 4;7. No grape has given rise to more controversies, suppositions, and investigations into its origin than this native red-wine variety. It is one of the oldest of the American varieties, and a sort of companion variety to Herbemont. Like Herbemont, it was taken to Europe during the phylloxera epidemic, and is there still grown to some extent for a very dark blending wine. Frequently encountered synonyms are Jacquez and Black Spanish. The vine is vigorous, but very sensitive to mildew; and the fruit is subject to both mildew and blackrot. In proper locations —that is, areas of low rainfall and long seasons—it is by all odds the best of the native red-wine hybrids.

Norton. See *Cynthiana* above.

FRENCH HYBRIDS—WHITE WINE

Couderc 299–35. Early midseason. True muscat aroma, suitable for sweet wine of the muscatel type, for sparkling wines of the Asti spumante type, or to contribute aroma to a dry-wine blend by using in small proportion. Satisfactory disease resistance and hardiness, and productive, but of only moderate vigor. Much more vigorous when grafted. Spur pruning.

Meynieu 6. Early midseason. A hybrid from Bordeaux showing promise as a rugged and steady producer of good ordinary white wine. High sugar. Rustic and productive.

Galibert 261–13. Late midseason. A Sémillon hybrid from the south of France. Handsome large-berried bunches, but sets raggedly at Boordy Vineyard. Good vigor of vine. Best suited to a dry climate, such as the Southwest. Spur pruning.

Ravat 6. Midseason. A Chardonnay hybrid producing fine white wine of the white Burgundy type. Only moderate vigor, not fully hardy, but a heavy producer. Preferably grafted. Requires careful protection against powdery mildew. Considered the best as to quality of the white-wine hybrids. Spur pruning.

Ravat 34. Early. Another Chardonnay hybrid but without the full aroma of R. 6. Moderate vigor, satisfactory resistance and hardiness, and a satisfactory producer on spur pruning.

Ravat 51. Early midseason. Much more vigorous and hardy than the two other Ravats but because of its smaller bunches not so heavy a producer. Clean, crisp white wine recalling Chablis, less "moelleux" than Ravat 6. Suitable for champagne. Cane pruning.

Ravat-Tissier 578. Earliest of the superior white wine hybrids. A rather weak grower and subject to some winter kill-back. Spur pruning.

Seibel 4986, syn., Rayon d'Or. (S.405 × S.2007.) Early. Districts 2;3; warmer parts of 6; cooler parts of 7. Vigorous vine except when planted on humid land, and resistant to the fungus diseases. Produces good crops when pruned to short canes and spurs; and it has the virtue of producing a partial crop following a late killing spring frost. Fruit clusters small to medium, very compact, berry oblate. Fruit easily develops 20–21% sugar, and the resulting wine is of good quality, in France being considered one of the best of the hybrid white wines. It is grown a good deal in the Loire Valley. A reliable vine, which appears adapted to a diversity of conditions. Must be picked promptly when ripe.

Seibel 5279, syn., L'Aurore. (S.788 × S.29.) Very early. Districts 1;2;5(?);6;8. Vine of above-average vigor and of drooping rather than upright growth, which ought to be pruned to four canes with spurs. It is a healthy vine. Clusters are long, slender, loose, sometimes shouldered. Berries round, grayish yellow, below average size, with attractive spicy flavor and liquid pulp, making crushing and pressing very easy, with a high yield of juice. The wine has a special and easily recognized aroma, is well balanced, and usually contains a trace of residual sugar. In the grape-growing regions of Switzerland it has a special use – namely, for freshly pressed unfermented grape juice which is sold in the cafés at vintage time. A grape of high promise for this country, especially for regions with short growing seasons. It is rapidly becoming a standard variety for the commercial production of white wines in the New York Finger Lakes district.

Seibel 9110. Early midseason. Districts 3; cooler parts of

4;7. Vine of average vigor, excellent health, with handsome, rather delicate, deeply serrated foliage, which turns bright yellow, like maple leaves, in the fall. Cluster is above average in size, rather loose, shouldered, very good to look at. Berries are egg-shaped, yellow gold in color, of medium size, with tender adherent skin, crisp texture and exquisite flavor. This variety makes delicious table grapes, far superior either to the standard table grapes from California or to the standard native table-grape varieties. In addition, its wine is distinctly above average, with a delicate aroma, good balance, and attractive flavor. Fruit cracks sometimes when wet weather coincides with maturity. Bears average crops.

Seibel 10868. Early midseason. Has seen some extension in the Niagara peninsula of Canada but may not be sufficiently hardy for that area. Handsome, pale-pink fruit in small bunches, clean-flavored wine. Moderate vigor, sufficiently disease resistant. Short-cane pruning.

Seibel 13047. Early. Vigorous vine, disease-free, above-average production of a very neutral wine. Fruit borne in big open compound clusters. A good table grape too. Spur pruning.

Seyve-Villard 5276 Early. Medium vigor, very healthy and winter-hardy and a heavy producer of big, handsome compound bunches. Wine has excellent sugar-acid balance. In my view this grape will rival Seibel 5279 for first place among the white-wine hybrids in Districts 2, 3, 4. Good second crop after frost. Spur pruning.

Seyve-Villard 12309. Late midseason, for the South and Southwest. Heavy producer; big handsome compound bunches; fruit amber, turning to deep pink on the exposed side. Good wine and good to eat. Spur or short-cane pruning.

Seyve-Villard 12375. Midseason. First cousin to S.V. 12309 but ripens ten days earlier. Superbly healthy vine, big production of good wine. Very popular in southern France. Another variety of great promise in all but short-season areas. Spur pruning.

Seyve-Villard 14287. Early. Another true muscat, but small-berried, for muscatel or to heighten the aroma of neutral white wines. Moderate vigor, hardy except under extreme conditions, good production though bunches are often ragged. Spur pruning.

Seyve-Villard 23–410. Midseason. Offspring of S.V. 12375 with similar characteristics. Extremely handsome bunches. Spur pruning.

Vidal 256. Midseason. A hybrid of the grape called St. Émilion in the French cognac district, Ugni Blanc in southern France, and Trebbiano in the Chianti district of Italy, with fruit strongly resembling that parent and yielding wine of the same type. Very vigorous, good producer. Foliage almost always shows pockets of downy mildew, which, however, are self-healing. Spur or short-cane pruning, depending on vigor.

White Baco. Early. Great vigor and a good producer. Satisfactory disease resistance. Gives promise as a variety yielding good ordinary white wine with minimum care. Cane pruning.

FRENCH HYBRIDS—RED WINE

Baco No. 1. (Rip. × Folle Blanche) Early. Districts 1(?);2;3;6(?). Vine of prodigious vigor, with huge leaves resembling those of its *riparia* parent. It is practically immune to the mildew and blackrot, save for an occasional touch of mildew at the apex of the bunch. The small berries are borne in long, loose clusters, usually shouldered. The fruit normally shows 20–21 per cent sugar, and acidity a bit above the average. Crushes easily and gives a high yield of must. Must always ferments rapidly and cleanly and the young wine is bright in a minimum of time. The wine has a clean flavor, an attractive slightly spicy but non-foxy aroma, and adequate but not heavy color. It gains quality with aging. Since the vines are so vigorous, they ought to be planted at least eight feet apart in the row. Trained to either the Kniffin or the cordon system it gives above-average crops.

Burdin 7705. Early midseason. A hybrid of the Beaujolais Gamay, or Gamay *à jus blanc*, which shows the fruit character of this parent and yields wine of distinct Beaujolais type. Hardy, good vigor, and satisfactory disease resistance. A moderate producer. Short-cane pruning.

Couderc 7120. Late midseason. One of the first of the successful French hybrids, it is still widely grown in southern France. Very vigorous and productive, good *ordinaire.* It definitely prefers hot, dry conditions.

Foch (Kuhlmann 188–2). One of a group of early-ripening Alsatian red-wine hybrids of *riparia* by Pinot Noir and Gamay. Adapted to short-season areas such as northeastern and north-central states. Just as Baco develops bouquet resembling claret, so Foch when well made somewhat recalls Burgundy though without the finesse. Vigorous and healthy. Cane pruning.

Galibert 115–24. Midseason. Does best in hot and relatively dry climates yielding Mediterranean-type wine. Sufficiently hardy and disease resistant. Spur pruning.

Joannes-Seyve 26–205. Mid-season. Vigorous, hardy, good production of big bunches, practically disease-free. Yields a wine without pronounced aroma but with the *sève* or body that many of the hybrid wines lack. In brief, thoroughly French. Currently enjoying a vogue in France, especially in the Loire Valley and the Touraine. Spur pruning.

Landot 244. Early midseason. Though its parentage contains no Gamay, L. 244 yields a deep-colored and remarkably Beaujolais-like wine. One of the best for quality. Moderate producer only, but healthy. Short-cane pruning.

Landot 4511. Midseason. This is Landot 244 × S.V. 12375, designed to combine the high red-wine quality of the former with the great vigor and productivity of the latter. It is vigorous, disease-resistant, and a heavy producer. Good wine but not quite equal in quality to L. 244. Spur or short-cane pruning.

Millot (Kuhlmann 192–2). Early, about a week ahead of its cousin Foch. More vigorous and a somewhat heavier producer in Boordy Vineyard than Foch. Wine has deeper color and seems superior. Cane pruning. Another even earlier variety in this series is *Joffre,* which is suitable for extremely short-season areas. Its bunches are smaller, its wine simply a good *ordinaire.* It ripens too early for the Middle Atlantic States and the birds get all the fruit.

Ravat 262. An early midseason Pinot Noir hybrid. Somewhat irregular in bearing and a bit subject to spray burn of the foliage, but worth a trial for the quality of its wine. Short-cane pruning.

Seibel 2, 14, 128, 1000, 4643, 5437, 6339, 6905, 10096. Entered for the record, veterans of the early days of hybrid growing in France and, years later, in this country. They have been superseded by later varieties that are superior in one respect or another. But any one of them yields a better red wine by European standards than any of our native red-wine varieties. Seibel 1000 is in fact being grown in New York and Ohio in substantial acreage; but it is capricious in its bearing habit, hence not eligible for much further extension.

Seibel 5898. Early midseason. One of the earlier Seibels, being grown to some extent in the Finger Lakes district of New York, primarily to contribute deep color to red-wine blends. As a *teinturier*, or color grape, it is inferior to S. 8357.

Seibel 7053. Early midseason. One of the two or three most widely grown hybrids in France, producing a well-balanced ordinary wine. It is a very heavy producer, every sucker and water sprout carrying flowers, so that it must be drastically pruned, especially in its early years, lest it bear itself to death. Always a good second crop after frost. From the Finger Lakes it is reported susceptible to powdery mildew. In Boordy Vineyard it has shown no damage from any of the three mildew diseases over two decades.

Seibel 8357. Midseason. Hardy, vigorous, and productive; no disease trouble. This is a *teinturier*, or color grape, intended to strengthen the color of light red wines and is used for that purpose throughout France. Its wine is not for separate use and is never added in quantities higher than 5 per cent. Its color is ten times the intensity of normal red wines. Cane pruning.

Seibel 10878. Midseason. Great vigor, good average production of wine slightly recalling the red burgundies. Its bud eyes push later than most, an advantage in frosty spots. Requires careful protection against the mildew diseases. There is a large acreage of this in the Canadian Niagara peninsula. Cane pruning.

Seibel 13053. Early. Very hardy and resistant and one of the best for short-season areas. Good producer of superior wine, rather neutral in flavor and light in color, which blends well with Baco and Foch. Also makes a good rosé. There is substantial commercial acreage of this in the Finger Lakes.

Seibel 13666. Early midseason. Average producer of superior deeply colored wine, satisfactorily disease resistant but only moderately vigorous and hardy. Short-cane pruning.

Seibel 14596. Midseason. Extremely vigorous; heavy producer of huge bunches, producing a good ordinary wine. Promises well for the Southwest and other long-season areas. Spur or short-cane pruning.

Seyve-Villard 5247. Early. Very vigorous vine of stocky growth, hardy and resistant, huge compound bunches of light blue fruit. We have found it one of the steadiest producers, year in and year out, of a light-bodied red wine of good quality and of a still better rosé. Cane pruning.

Seyve-Villard 18–283. Late. Extremely vigorous, rangy vine producing large crops of huge bunches. Satisfactory hardiness and resistance. Wine of good ordinary quality. Spur pruning.

Seyve-Villard 18–315. Late. Vigorous and healthy, much grown in southern France, and almost as heavy a producer as S. 7053. Its wine is deep-colored, somewhat astringent, excellent for blending. Spur pruning.

Seyve-Villard 20–347. Late. A French Midi grape,

adaptable to our Southwest. Heavy producer, can do double duty as a table grape.

Seyve Villard 23–657. Late. Grown a good deal in the lower Rhone Valley of France. Wine heavy-bodied, high in alcohol. A reliable producer in hot, dry areas. Spur pruning.

VINIFERA IN THE EAST

After all I have written about sad experiences over three centuries with *vinifera* grapes in the East—that is, outside of California and its viticultural extensions[1]—it may seem strange to return to the matter. Yet the question has been reopened. For a decade it has been the subject of renewed experiment and of lively, sometimes acrimonious, discussion among eastern wine-growers.

Two men are responsible for this renewal of interest, Dr. Konstantin Frank and Charles Fournier, the latter of Gold Seal Vineyards, one of the prominent wineries in the Finger Lakes district of New York.

Dr. Frank is a native of one of those enclaves of German language, culture, and race which so inappropriately dot the western part of Soviet Russia. His early professional life was spent growing the hardier *viniferas* under the difficult conditions of the Ukraine with its continental extremes of climate (some excellent white wines are grown in the Ukraine). When he reached this country not long after the war, he gravitated naturally to the Finger Lakes district. There he reached the conclusion that if the *vinifera* would grow in the Ukraine they could be made to grow in what he considered the no more rigorous climate of central New York. His proposition was greeted with general skepticism. But in Mr. Fournier, French-born with a background in the French champagne industry, he found the necessary early support for his experiments.

Dr. Frank's argument was essentially this: that the *vinifera* can survive low winter temperatures (and in the Finger Lakes they occasionally drop as low as −24°F)

[1] See page 58 for these extensions.

provided the vines go into their dormant period fully ripened. This means that they must be meticulously protected during the growing season against disease, root disease and above-ground disease, so that the physiology of the vine is not interfered with. It also means that early ripening of the woody parts of the vine must be induced by suitable rootstocks – rootstocks that not only have good affinity with the scion but have a short growth cycle.

Since efforts to grow *vinifera* in the East were abandoned years ago, there had been great strides in the development of better fungicides and insecticides and in knowledge of their use. This was an important advantage. As to rootstocks, Dr. Frank proposed to try the more promising rootstocks already in existence, and also to search out and try native wild vines of upper New York and, still farther north in the St. Lawrence Valley, those of the species *riparia.* This is extremely hardy, roots well and takes grafts well, is highly resistant to phylloxera, and has a short cycle of growth.

To cut the story short, there have been both successes and failures. The failures have been instructive. They show that soil drainage and the right rootstock are vital. They show that a great many of the *vinifera* cannot take it. This seems to apply more to varieties from the hot-country ranges of *vinifera* territory such as the Mediterranean basin than to the more northerly sorts. They also suggest that the red-wine *vinifera* are in general harder to bring through than those for white wine. And as might be expected, a good many of the standard rootstocks are not suitable.

The successes have been encouraging. Some of the more northerly, short-season *viniferas* on certain rootstocks survived even the dreadful northeast winters of 1961–2 and 1962–3 (a succession of bad winters is the severest test) and bore crops the following season. Several of these varieties have been planted in modest commercial quantities and their wines have found a ready market at premium prices. They have been in every way comparable to superior white Burgundies and Rhine wines.

A fair statement of the present situation includes the following points. These *vinifera* are exotics, as the natives and the French hybrids are not. The care of them requires expertise of a high order. Only a limited number can yet be considered reliable, even with expert care. It seems unlikely that they will ever become workhorses of the eastern vineyards, but for growers who accept the risk they may find a place as supplements for the production of especially fine wines. On this basis, the effort deserves encouragement and broader experiment, perhaps best carried out by small producers and qualified amateurs who really understand the risk. The list of varities which can be given a qualified recommendation at this stage of the work is small, but the wine quality of all of them is of course high.

Aligoté. Very widely grown for white wine in central Europe and the Russian wine-growing districts as well as in Burgundy. Heavy producer.

Cabernet Sauvignon. Some Cabernet has been produced in New York state, but it is not particularly impressive; and behavior of the vine is in general less satisfactory than the whites. Here in Boordy Vineyard, Ruby Cabernet from California (Carignane x Cabernet) occasionally yields a little wine; Cabernet has not.

Chardonnay. The white Burgundy grape *par excellence.* Appears to be the hardiest of the lot and least subject to disease, and at Boordy Vineyard produces a modest quantity of fine wine; one of the two most promising in the Finger Lakes.

Muscat Ottonel. A true muscat with high aroma and flavor, and one of the more promising culturally.

Pinot Gris. A pink sport of Pinot Noir, for white wine. Quite hardy, good producer. This grape is grown a good deal in Alsace.

Pinot Noir. The grape from which fine red Burgundies are made; wines made from the grape in vineyards alongside Lake Keuka have been fully characteristic. A rather shy producer.

Riesling. The leading white-wine grape of the Rhine, the Moselle, and Alsace. Disease has so far been very troublesome as far south as Maryland. It appears clean and healthy in the Finger Lakes, and on certain rootstocks it has grown and borne well along Lake Keuka in the Frank and Gold Seal vineyards. Several famous clones have been tried and seem just about on a par culturally.

Traminer. Grown in the Finger Lakes, Traminer has all the fragrance one associates with the Alsatian Traminers. But it appears less hardy than either Riesling or Chardonnay.

Riesling and Chardonnay have so far been the stars of this slow and continuing experiment; and if two such grapes prove practical in parts of the East, the effort will have been more than justified. But no final conclusion on the *vinifera* can yet be drawn. These few paragraphs should be taken merely as a progress report. On the question of rootstocks, see pages 195–7.

BIBLIOGRAPHY

I have boiled this down to a few basic things, in effect a "course" in wine-growing. Taking off from these, determined students will find a huge field for browsing.

AMERINE, M. A., and WINKLER, A. J.: "Composition and Quality of Musts and Wines of California Grapes." *Hilgardia* XV:6, 1944. [Basic study that has guided the choice of wine varieties since its publication.]

CATO, MARCUS PORCIUS: *De Agri Cultura.*

CHANCRIN, E.: *Viticulture Moderne.* Paris. [Standard French handbook.]

DANIEL, LUCIEN: *La Question Phylloxérique.* 3 vol. Bordeaux, 1908, 1910, 1915.

GALET, P.: *Cépages et Vignobles de France.* Tomes I–IV. Montpellier, 1956–64. [Tome I contains a thorough and highly critical review of the French hybrids.]

HEDRICK, U. P.: *The Grapes of New York.* Albany, 1908. [Classic description of the older American hybrids in quarto with dozens of color plates.]

JACQUELIN, L., and POULAIN, RENÉ: *The Wines and Vineyards of France.* New York, 1962. [Handsome and detailed description of French vineyards and crus, region by region.]

LANGENBACH, A.: *German Wines and Vines.* London, 1962.

LEVADOUX, LOUIS: *La Vigne et Sa Culture.* Paris, 1961.

MENDALL, SEATON C.: *Vineyard Practices for Finger Lakes Growers, The Planting and Care of Young Vineyards in the Finger Lakes Area,* and *Grape Disease and Insect Control Guide for Finger Lakes Counties.* Hammondsport, 1957, 1959, 1964. [Detailed instructions for growers supplying wine grapes to the Taylor Wine Company.]

MUNSON, T. V.: *Foundations of American Grape Culture.* Denison, Texas, 1909. [Basic work on the American species and on Munson's own hybrids.]

Bibliography

PEROLD, A. I.: *A Treatise on Viticulture*. London, 1927. [From the South African point of view.]

SCHOONMAKER, FRANK: *Encyclopaedia of Wine*. New York, 1964.

SHAULIS, N., and JORDAN, T. D.: *Cultural Practices for New York Vineyards*, and *Chemical Control of Weeds in New York Vineyards*. Cornell Extension Bulletins 805 and 1026.

VARRO, MARCUS TERENTIUS: *On Agriculture*. [A good translation will be found with a translation of Cato, in the Loeb Classical Library.]

VIALA ET VERMOREL: *Traité Général de Viticulture*. 7 vols. Paris, 1901–09. [A series of thorough monographs, each by a specialist, on the classic varieties of *vinifera*.]

WAGNER, PHILIP M.: *American Wines and Wine-making*. 5th edition. New York, 1963.

WINKLER, A. J.: *General Viticulture*. Berkeley, Calif., 1962. [Exhaustive study, with the emphasis on California viticulture.]

INDEX

(Names of grape species and varieties are printed in *italics*. The principal page reference to each grape variety is likewise printed in *italics*.)

Index

Index

INDEX

Index

Index

Index

A NOTE ON THE TYPE IN WHICH THIS BOOK IS SET

This book was set on the Linotype in Janson, a recutting made direct from the type cast from matrices made by Anton Janson some time between 1660 and 1687.

Of Janson's origin nothing is known. He may have been a relative of Justus Janson, a printer of Danish birth who practised in Leipzig from 1614 to 1635. Some time between 1657 and 1668 Anton Janson, a punch-cutter and type-founder, bought from the Leipzig printer Johann Erich Hahn the type-foundry which had formerly been a part of the printing house of M. Friedrich Lankisch. Janson's types were first shown in a specimen sheet issued at Leipzig about 1675. Janson's successor, and perhaps his son-in-law, Johann Karl Edling, issued a specimen sheet of Janson types in 1689. His heirs sold the Janson matrices in Holland to Wolffgang Dietrich Erhardt.